Electron-Ion-Plasma Modification of a Hypoeutectoid Al-Si Alloy

Electron-Ion-Plasma Modification of a Hypoeutectoid Al-Si Alloy

Dmitrii Zaguliaev • Victor Gromov
Sergey Konovalov • Yurii Ivanov

CISP

CRC Press
Taylor & Francis Group
Boca Raton London New York

CRC Press is an imprint of the
Taylor & Francis Group, an **informa** business

Translated from Russian by V.E. Riecansky

CRC Press
Taylor & Francis Group
6000 Broken Sound Parkway NW, Suite 300
Boca Raton, FL 33487-2742

© 2021 by CISP
CRC Press is an imprint of Taylor & Francis Group, an Informa business

No claim to original U.S. Government works

Printed on acid-free paper

International Standard Book Number-13: 978-0-367-49380-6 (Hardback)
International Standard Book Number-13: 978-0-367-49386-8 (Paperback)

Visit the Taylor & Francis Web site at
http://www.taylorandfrancis.com

and the CRC Press Web site at
http://www.crcpress.com

Contents

Introduction

The problem of destruction of parts under the influence of external mechanical loads has long been known. The traditional method of solving this problem is the volumetric modification of products. However, to increase the service life of the product, in most cases, hardening of its surface layers is sufficient, due to the fact that their destruction begins with the surface. This is due to the fact that the most intense plastic deformation occurs in the surface layers, with a depth of grain size [1]. Improvement of properties can be achieved by treating the surface with concentrated energy flows, such as laser radiation, powerful ion beams, plasma flows and jets [2–14]. These types of processing allow modification locally, that is, in those places where destruction occurs during the operation of the product.

The main feature of hardening materials with concentrated energy fluxes, in comparison with the methods of traditional thermal and chemical-thermal treatment, is the nanostructuring of their surface layers. This means a decrease in the scale level of localization of plastic deformation of the surface, which leads to a more uniform distribution of elastic stresses near it under the influence of operational factors. As a result, the probability of nucleation of microcracks in the surface layer leading to failure is significantly reduced. This increases both strength and ductility. This allows us to solve one of the most important problems – ensuring the optimal ratio of surface properties and material volume. In this case, there is no need to use volume-alloyed materials and it becomes possible to a certain extent to solve the important task of mechanical engineering – improving the reliability and durability of large parts in operating conditions.

The application of these surface treatment methods in various industries, for example in the space or aviation, is steadily increasing and becoming comparable with traditional coating methods.

One of the promising methods of external energy exposure, which has a significant effect on the structure, phase composition, physical and mechanical properties of the surface layers of metals and alloys, is multiphase plasma jet treatment. This method allows to obtain high-quality coatings with good adhesion to the substrate and high functional properties. The method allows to apply coatings from the products of explosion of conductors, as well as to form composite coatings significantly superior in their properties to the starting material [15,16], since more durable materials are usually used as the coating material, in this book we consider the coating using the yttrium oxide.

It is known that yttrium is a metal with a number of unique properties, and these properties largely determine its very widespread use in industry today and, probably, even wider application in the future. The tensile strength for undoped pure yttrium is about 300 MPa. A very important quality of both yttrium metal and a number of its alloys is the fact that, being chemically active, when heated in air, yttrium is coated with a film of oxide and nitride protecting it from further oxidation to 1000°C. Promising fields of application for yttrium alloys are the aerospace industry, nuclear technology, and the automotive industry.

The second of the most promising and highly effective methods of surface hardening of products is electron-beam processing [17]. Electron-beam processing provides ultrahigh heating rates (up to 10^6 K/s) of the surface layer to predetermined temperatures and cooling of the surface layer due to heat removal to the bulk of the material at speeds of 10^4–10^9 K/s, resulting in the formation of non-uniform submicron and nanocrystalline structural phase states [18].

This type of processing has wide possibilities of controlling the input energy, low energy reflection coefficients, and a high concentration of energy per unit volume of the material [19]. Electron beam processing has several advantages over other surface modification methods. Compared to high-power ion beams, electron-beam processing is generated with a significantly higher coefficient of efficiency in a pulse-frequency mode at lower accelerating voltages and does not require the creation of special radiation protection. High energy efficiency, high uniformity of energy density over the flow cross section, good susceptibility of pulses and a high repetition rate provide several advantages over the pulsed flow of low-temperature plasma [20–23].

The main advantage of electron beam processing is the combination of virtually complete absorption of electron energy with the possibility of varying the depth of electron penetration into the material, and, accordingly, the dynamics of thermal fields and stress wave parameters [24].

In some cases, a single-component exposure becomes insufficient, and then it is necessary to resort to complex modification (a combination of several methods of energy exposure) [25–27].

In the presented book, all three described methods of energy exposure are considered.

The monograph is concerned with theoretical and experimental research computer simulation of structural phase transformations in Al-11Si-2Cu alloy on different scale levels under electroexplosion alloying, electron beam processing and electron–plasma alloying proceeding at the nanolevel in order to create new materials. Six modes of processing were analyzed for each type of external energy effect and the optimal modes resulting in multiple increase in strength, durometric and tribological characteristics of 100 mm thick surface layer alloy were determined. It was revealed that enormous increase in the physical and mechanical properties of Al–11Si–2Cu alloy under electroexplosion alloying is reached at the expense of creation of a multilayer, multiphase nanocrystal structure formed largely by oxides and silicates of aluminium and yttrium in the surface layer. It was stated that electron beam processing resulted in the formation of a surface whose mechanical characteristics increased significantly the values of the alloy in the cast state which is reached due to the formation of a fine-grained, gradient, cellular structure free from intermetallides. Electron-ion-plasma alloying leads to the cardinal structural transformation of material's surface layer consisting in the formation of a multielemental multiphase layer having a submicrocrystalline structure. Dissolution of silicon inclusions and intermetallides of micron and submicron dimensions characteristic of the cast state alloy takes place.

The field of application of the results presented in the monograph is research, scientific and technological enterprises involved with modification of light alloy surfaces with the aim of their use in the automobile and aerospace industry.

The book is intended for specialists in the field of physical material science, condensed state physics, metal science and thermal treatment and it may be useful for undergraduate and post-graduate students of the corresponding specialties and directions of training.

The authors are grateful to their colleagues who took an active part in the research and discussion of the results included in the monograph – V.D. Sarychev, S.A. Nevskii. Special thanks are due to the reviewers V.A. Danilov and V.V. Maruv'ev for valuable comments, which were taken into account when preparing the manuscript for publication.

This work was supported by Russian Science Foundation [project No. 19-79-10059].

Analysis of modern methods of surface modification of light metals using external energy sources

1.1. Plasma spraying and electron beam surfacing of wear-resistant coatings

In engineering technologies, to provide the required level of wear resistance of machine parts, plasma spraying of wear-resistant coatings is widely used [28,29]. Among the advantages of the plasma spraying method, one can single out high productivity, good process control, as well as the ability to process parts of various configurations and dimensions.

In the study [30], high-quality Ti/TiBCN coatings were obtained on the surface of 7075 aluminium alloy by laser welding. The coating material was Ti powder (99.5% purity, crystallite size 100–150 µm) and TiBCN powder (98.5% purity, crystallite size 100–150 µm); in the entire series of experiments, the mass of Ti powder did not change, and only the TiBCN content varied (0 wt.%, 5 wt.%, 10 wt.% and 15 wt.%). After coating, the structure was analyzed using scanning electron and transmission microscopy. Microhardness was used as a parameter evaluating the strength characteristics of the resulting coatings. As a result of studies, it was found that the strength characteristics of coatings increase with increasing TiBCN content and reach maximum values at 15 wt.%. Microstructural analysis showed that the cross-sectional structure of the sample, after laser surfacing, is divided into a coating zone, a transition zone, a heat-affected zone and a substrate. Coatings mainly consist of equiaxed grains and white lattice-like crystals, and the transition zone mainly consists of elongated dendrite crystals and white small particles. Q. Wang and his science group [31] studied three different Fe-based powder materials, gray cast iron, high chromium steel, and self-fluxing powders with a high content of chromium and nickel. Coating was carried out by plasma spraying on an Al–Si system alloy substrate. The microstructure, hardness, phase composition, substrate adhesion, and wear resistance of the deposited coatings were studied. The sprayed coatings showed improved wear resistance compared to the original Al–Si alloy in terms of friction coefficient, mass and volume losses. The best integral characteristics were shown by chromium steel coatings. Various wear mechanisms have been identified: a mixture of adhesive and abrasive wear for gray cast iron coatings, oxidative dominant wear for chromium steel coatings, and a mixture of oxidative and fatigue wear for self-fluxing coatings.

Another example of such studies can be found in [32], in which the creation of a high-strength coating consisting of a mixture of aluminium powder with carbon nanotubes on the surface of an aluminium alloy is considered. It was found that, depending on the speed of laser processing, the mechanical characteristics of the final product change. The results showed that the microhardness of the coating was 43% higher than the microhardness of the substrate. The authors attribute such changes in the mechanical characteristics to the diffusion of alloying elements from the substrate into the melt pool, which was formed as a result of laser melting, and it is also worth noting the obvious contribution of carbon nanotubes to the microhardness of the modified layer.

In addition to aluminium, magnesium and alloys based on it are actively used in the modern aerospace and automotive industries, due to its high specific strength and stiffness. However, the low corrosion resistance of magnesium and magnesium alloys significantly limits their widespread use. Magnesium and its alloys can be protected by forming protective surface layers, which can be achieved by plasma spraying, capable of producing metal and / or ceramic coatings. In [33], NiAl10 and NiAl40 plasma coatings on a AZ91 magnesium alloy substrate were obtained using a hybrid plasma spraying system. The results show a significant effect of the preheating temperature of the AZ91 substrate during plasma sputtering on the development of diffusion bonding due to the formation of a sublayer consisting of the Mg3AlNi2, Al12Mg17 phases and Mg and Al solid solutions.

Potentiodynamic measurements showed a twelve-fold increase in the polarization resistance (930 W cm²) of the NiAl10 coating compared to the initial alloy AZ91, the NiAl40 coating showed an almost twofold increase in polarization resistance (112 W cm²). Long-term corrosion tests showed a significant positive effect of the sublayer formed from the eutectic structure of the Al12Mg17 phase with a solid solution of magnesium and aluminium on the corrosion resistance of NiAl40 plasma coatings.

In [34], the authors studied coatings formed on VT6 titanium alloy by plasma spraying of alumina powder, grade 25AF230 and subsequent microarc oxidation (MAO) at different current densities. The current density during MAO influenced the morphology of the coating and the average size of the structural elements of the coating. Open porosity in the form of open pores, as well as cracks, decreased from 50.3 ± 4.5% to 10.3 ± 1.5%. As a result of subsequent oxidation, the microhardness increased by 80–115 HV2 depending on the current density. Microarc oxidation also allows the formation of structural elements in the form of pores and crystals from 15–150 μm in size on the surface of the deposited alumina coatings.

Non-vacuum electron beam surfacing is an effective method for obtaining corrosion-resistant coatings of the Ti–Ta–Nb system formed on plates of technically pure titanium grade VT1-0 [35]. As a result, it was found by optical and scanning electron microscopy that the coating has a complex structure formed as a result of nonequilibrium cooling of the melt. A number of characteristic zones can be distinguished in the coating structure: 1. zone of deposited metal; 2. base metal zone; 3. transition zone. In the coating structure at different scale levels, traces of the dendritic structure, grain and subgrain boundaries, and hardened areas with a needle structure can be distinguished. The corrosion resistance of the obtained coating significantly exceeds the corrosion resistance of the VT1-0 alloy ≈ 6 times.

The authors of [36] proposed a technology for the formation of multilayer corrosion-resistant tantalum-containing coatings by the method of non-vacuum electron beam surfacing on the surface of workpieces made of technically pure VT1-0 titanium. The technology allows forming defect-free coatings up to 4.5 mm thick. Multilayer coatings have a complex gradient structure, both in the cross section of the deposited layer and in the cross section of the composition as a whole. The formation of tantalum-containing coatings on titanium leads to hardening of the material, which is confirmed by the results of durometric studies. An increase in the strength characteristics of surface-alloyed layers has a negative effect on the level of impact toughness of the material, which decreases by 35–45% compared with technically pure titanium. The level of corrosion resistance of two-, three-, and four-layer coatings in boiling 68% nitric acid is two orders of magnitude higher than the level of corrosion resistance of technically pure titanium in a similar medium and is only 6–7 times lower than that of pure tantalum.

The use of aluminium alloys in friction units is a promising task in mechanical engineering. Aluminium alloys have advantages over many structural materials: ease of processing, low density, acceptable strength characteristics. However, their low resistance to mechanical wear does not allow them to be applied without surface modification. In [37], the issues of increasing the wear resistance of the drive variator disks of a combine harvester made of AK9 aluminium alloy by applying a plasma coating were considered. For the experiment, samples were made of cast iron SCh18-36 and aluminium alloy AK9. A plasma coating of PN85Y15 powder 0.5...0.7 mm thick was applied to samples from AK9 aluminium alloy. To determine the comparative wear resistance of the samples, an accelerated test method was used, in which a flat sample is abraded with a reference disk with the abrasive fed into the friction zone. As a result of the tests, the PN85Yu15 plasma coating on AK9 aluminium alloy is 3.7 times superior in wear resistance to SCh18-36 cast iron.

A known method of chemical-thermal hardening of the surface of titanium alloys and products [38] includes heating the surface of the product in a nitrogen environment, the heating is carried out by a concentrated heat source with a power density of 10^3–10^4 W/ cm^2, current 80–150 A and the speed of the source relative to the product of 0.005–0.01 m/s The concentrated heat source was an electric arc or a plasma jet. The result: increased wear resistance and corrosion resistance of parts made of titanium alloys.

A method is described in [39] for the manufacture of a catalytic composite coating, which comprises producing a catalytically active layer by plasma spraying. Before applying the catalytically active layer, an adhesive layer is applied by spraying a powder composition containing, wt.%: aluminium 3–10, aluminium hydroxide the rest, and the subsequent catalytically active layer is applied by a powder composition containing, wt.%: aluminium 3–5, chromium 2–5, tungsten oxide 0.8–1.2, oxides of cerium, lanthanum, neodymium in the amount of 1.8–2.2, copper oxide 2–3, aluminium hydroxide, the rest. Then, using an ion-plasma method with two evaporators, an activator layer is applied containing, wt.%: copper oxide 27-34, chromium oxide 66–73. The powder composition is applied at a distance of 15–50 mm from the substrate. The thickness of the catalytically active layer is set in the range of 30–100 µm, the thickness of the third layer of activator is set in the range of 4–6 µm.

A method [40] of hardening a cutting tool, including the deposition of a multilayer coating of the Ti–Al system is available. The cutting tool is placed in the working chamber on the table, the activation of its surface before deposition of the multilayer intermetallic coating of the Ti–Al system is carried out by heating and cleaning the surface using a plasma source of a filament cathode and electric arc evaporators. A multilayer coating is applied while simultaneously spraying two single-component cathodes of Al and Ti and rotating the table around its axis with TiAlN and TiAl layer-by-layer, with argon used for sputtering titanium–aluminium and nitrogen for sputtering titanium aluminium nitride as a working gas. Gas change is carried out using a gas flow regulator, while layer-by-layer coatings are sprayed in a single cycle with alternating deposition of TiAlN and TiAl layers, which are repeated at least 10 times, while the table rotational speed is 10 rpm, the thickness of each layer is from 5...50 nm with a total coating thickness of up to 5 µm, in which TiAl3, Ti3Al intermetallic phases are formed, in the pure form of Ti and Al.

The method described in [41] is used for producing a coating based on complex nitrides, in which a substrate is placed in a vacuum chamber of a facility equipped with magnetron sprays, electric arc evaporators and a resistive heater, the surface of the substrate is cleaned in a glow discharge when the surface is contactless heated by a resistive heater to 100°C and ion cleaning by an electric arc evaporator in an inert gas medium when the surface is heated to a temperature of 300...350°C, then a lower layer of titanium and alternating layers of nitride are applied to the substrate in a mixture of inert gas and nitrogen.

In the method [42] for treating a TiC–Mo system with a multiphase plasma jet of a composite wear-resistant coating on a friction surface, a titanium carbide powder sample is placed between two layers of molybdenum foil, an electric explosion of the foil is carried out with the formation of a pulsed multiphase plasma jet, and its friction surface is melted with a specific energy flux of 3.5...4.5 GW/m^2 and sputtering on a fused layer of plasma jet components with subsequent self-hardening and the formation of composition coating containing titanium carbide and molybdenum.

A method has been developed [43] for depositing heat-resistant and wear-resistant coatings based on titanium aluminides on titanium and titanium alloys, including conducting electric arc welding with a non-consumable electrode in inert shielding gases using a filler wire. An aluminium wire is used as a filler wire, and surfacing is carried out in modes that provide a deposited layer with an aluminium content of 5–25%.

A known method [44] is the deposition of alloys based on titanium–copper intermetallic compounds on titanium and titanium alloys, including arc welding by a non-consumable electrode in inert shielding gases using a filler wire. The filler wire is made of copper or copper alloys, and surfacing is carried out in modes that provide a deposited layer with a copper content of 5–40%.

1.2. High-intensity electron and powerful ion beams, laser radiation

The literature has written in detail about the flaws of the surface properties of aluminium alloys, which seriously limit their further application in many fields [45]. The main problem is the low resistance to localized effects, in particular pitting corrosion, caused by the destruction of an oxide film exposed to the atmosphere, fresh or salt water, and other electrolytes. To overcome this drawback and increase the stability of the oxide film, surface alloying of aluminium with such transition metals as Mo, Cr, W, and Ta is performed [46, 47]. But since the solubility of these components in aluminium is less than 1 at.% various methods of exposure to concentrated energy flows are used, one of which is a pulsed electron beam. Irradiation by a pulsed electron beam gives rise to huge inhomogeneities in the distribution of temperature fields in the surface layers of the material and leads to ultrafast melting, mixing, and high-speed crystallization. In [48], surface alloying with molybdenum of an aluminium alloy by a high-current electron beam was investigated. As a result, it was found that, after irradiation, an Al5Mo phase with a acicular structure appeared in the doping layer. Numerous structural defects have been discovered, such as craters, various cracks, dislocation loops, and dislocation walls. Studies of various irradiation regimes showed that with an increase in the number of pulses, the density and size of the craters formed on the irradiated surface decreased significantly, and a large increase in corrosion resistance was also observed. The international scientific team investigated the influence of the electron beam scanning speed on the surface of the Al–3Ti–1Sc aluminium alloy [49]. Experimental studies were carried out in a vacuum chamber in five modes, differing in surface scanning speeds. Five passes were made at speeds of 3, 8, 12, 15 and 20 mm/s. The research results showed that the cooling rate, which is a function of the scanning speed of the electron beam, plays a key role in determining the microstructure of the Al–3Ti–1Sc alloy. At a slow scanning speed of 3 mm/s, an accelerating voltage of 50 kV and a beam current of 30 μA, the primary phases Al3(Ti, Sc, Fe) with a tetragonal lattice are still formed in the re-solidified melt, but they are much smaller in size (most of them are smaller than 2 μm) than the primary phases Al3(Ti, Sc) (~100 μm). When the scanning speed is increased to 20 mm/s, the primary Al3(Ti, Sc) phases are completely suppressed, and the newly solidified melt has a homogenized structure. It is also worth noting that the microhardness of the layer formed after scanning exceeds the microhardness of the initial alloy by 48%.

In recent years, numerous methods have been applied to modify the microstructure of aluminium alloys by laser surface hardening in order to increase their mechanical properties [50, 51]. This is an innovative hardening technology, which allows introducing high-amplitude compressive stresses and grinding the microstructure in the surface layers of the processed materials, which leads to an increase in mechanical properties [52]. In [53], the effect of laser processing on the microstructure and mechanical properties of the 2024-T351 aluminium alloy was studied; laser processing was performed at room and cryogenic (from –196°C to –130°C) temperatures. The research results showed that the surface microhardness, residual compressive stresses, and tensile strength of the laser-treated sample increased by 22.84%, 36.81%, and 11.88%, respectively. The authors also note that the use of laser processing in the cryogenic temperature region further increases the mechanical properties of the material, so the relative elongation, when tested for tensile testing, increased by 7.51% compared with the sample subjected to laser processing at room temperature. The authors explain these results of mechanical tests by the fact that the use of cryogenic temperatures during the process can effectively suppress slip and annihilation of dislocations, which leads to an increase in their density, thereby contributing to grain refinement and an increase in the mechanical characteristics of 2024-T351 aluminium alloy. As already mentioned, aluminium and its alloys are widely used in almost all areas of modern industry, including those where there are problems of cavitation wear of parts. Cavitation erosion is a phenomenon of destruction of the surface of a part, arising as a result of the rapid formation, growth and collapse of bubbles in liquids due to strong pressure fluctuations [54]. Shock waves repeatedly hit the surface of the part, which leads to the formation of pits, plastic deformation, strain hardening, the appearance and propagation of cracks, and ultimately the mass removal of the surface material. Z. Tong and the rest. [55] investigated the effect of laser shock hardening on the cavitation-erosion properties of AA5083 aluminium alloy. It was found that impact laser treatment leads to the formation of a fine-grained structure on the surface of the AA5083 alloy. The high density of the dislocation structure induced by laser impact treatment leads to an increase not only in strength but also in the ductility of the material. Along with the strength characteristics, cavitation-erosion resistance is also increased by 2.13 times compared with the untreated sample. The authors attribute these effects to thinning of grains and an increase in the density of dislocations during laser processing. A detailed analysis of the mechanical properties and microstructural evolution of a 2A14 aluminium alloy subjected to multiple laser impact hardening (LIH) was performed [56]. When analyzing the mechanical properties, tensile tests and the results of microhardness measurements were used, and the characteristics of the microstructure were studied out using scanning electron microscopy and transmission electron microscopy. Experimental results showed that the strength characteristics of 2A14 aluminium alloy improve with an increase in the number of impact laser processing cycles. After 3 cycles, the tensile strength and surface microhardness reached 525 MPa and 262 HV, respectively, which is 20.69% and 72.37% compared with the untreated sample. In addition, a high density of dislocation structures was found in the surface layer of the sample; this explains the serious increase in mechanical characteristics.

Laser heat treatment refers to modern methods of increasing the physicomechanical properties of the surface of machine parts. Compared with other sources of heating, the laser beam has a number of significant advantages, such as significant radiation power, exposure locality, a small heat-affected zone, the ability to process the surface of parts in hard-to-reach places and a high degree of automation of the processing process. New opportunities open up in the process of applying laser irradiation, when a combination of a high level of operational properties with the ductility of the product base is ensured. In this regard, the aim of research [57–59] is to study the physicomechanical properties of the surface layer of titanium samples after exposure to continuous laser radiation and to identify the optimal parameters of heat treatment, leading to a significant increase in microhardness and the formation of stable structures. Samples of technically pure titanium VT1-0 were processed, processed according to the scheme: pretreatment + annealing + continuous exposure to laser radiation. As a result, it was shown that the maximum value of Knupp microhardness occurs at the maximum speed of the laser beam V_{las} = 5 mm/s and is 900 NK compared to the initial value of 450 NK, which is explained by a significant cooling rate. However, such a regime leads to the formation of unstable, nonequilibrium structures, which was confirmed using x-ray phase structural analysis. Thus, from the point of view of increasing microhardness (Knupp microhardness increases to 850 units) and obtaining a calm structure, the regime with a laser beam velocity V_{las} = 4 mm/s is the result. Due to the increased strength characteristics, corrosion resistance, high biocompatibility, low thermal conductivity, titanium, zirconium and alloys based on them are widely used in chemical, machine-building, instrument-making, medical equipment and other industries both in Russia and abroad. However, the main disadvantage of titanium and zirconium are the low mechanical properties of the surface, for example, endurance under conditions of exposure to the product of periodic loads. In this regard, the aim of [60] was to study the process and comparative analysis of the effect of laser hardening of titanium alloy VT6 and zirconium grade E110 on the strength characteristics of their surface. It was found that pulsed laser radiation increases the microhardness of the treated surface of titanium samples to 14 ± 0.1 GPa and zirconium to 19 ± 0.1 GPa, as well as to the formation of a hardened surface layer with a thickness of up to 55 μm for titanium samples and a thickness of up to 30 μm for zirconium samples.

Currently, there is a growing interest in the use of aluminium alloys as a material for cylinder blocks and parts of the connecting rod and piston group of gasoline and diesel internal combustion engines. Compared to the traditionally used gray cast iron cylinder blocks, aluminium alloy blocks have a number of advantages: in addition to their low specific gravity, they have a high specific elastic modulus and good thermal conductivity, which provides significant unloading of thermally loaded zones. Due to the lower mass of the cylinder blocks and the parts of the connecting rod and piston group, the consumption of fuel and, accordingly, the emission of harmful substances are reduced. The authors of [61] studied the microstructure, microhardness, surface roughness, and chemical composition of cast aluminium alloy AK7ch after surface laser heat treatment using new absorbent coatings based on dextrin of compositions 1 (aqueous solution of dextrin + Na_2O $(SiO_2)n$) and 2 (aqueous dextrin solution + ZnO). Laser thermal treatment of the surface of the samples with deposited absorbent coatings with surface melting was carried out on a CO_2 laser with a radiation wavelength of $\lambda = 10.6$ μm with a radiation power of $P = 5.0$ kW and a diameter of a laser spot on the surface $d = 7$ mm and a fiber optic laser LS-10 continuous operation with a radiation wavelength $\lambda = 1.070$ μm with a radiation power of $P = 5.0$ kW and a laser spot diameter on the surface $d = 6$ mm. It has been established that the use of absorbing coatings based on dextrin leads to the formation of melted layers of considerable depth on the surface of the AK7ch aluminium alloy, when processing compositions 1 and 2 with a CO_2 laser, the depth of the fusion zone is 1.16 and 0.80 mm, respectively, and when processing using a fiber optic laser, respectively, 0.75 and 0.55 mm In this case, depending on the coating composition, the depth of the reflow zone during processing by radiation of a CO_2 laser is 1.45-1.55 times greater than when processing by radiation of a fiber-optic laser, i.e., coatings of both compositions are more effective when processing an aluminium alloy by the radiation of CO_2 laser. As a result of laser heat treatment, a substantial refinement of the structure and some hardening of the AK7ch alloy in the reflow zone occur. In this case, the size of α-Al dendritic cells decreases from 50–190 μm to 5.0–11.5 μm, the size of silicon crystals decreases from 5–30 μm to 0.5–2.0 μm, and the microhardness increases by 1.11 −1.22 times compared with the microhardness of the alloy in the initial cast state (90 HV0.025). The composition of the absorbing coating does not significantly affect the structure and microhardness of the AK7ch alloy after laser treatment. The study of the surface topography showed that when processing with a CO_2 laser, the surface roughness of the reflow zone is 3.60 (for coating composition 1) and 5.90 μm (for coating composition 2), respectively, and when processing with a fiber laser 1.03 and 3.90 μm. Moreover, depending on the coating composition, the surface roughness of the reflow zone during processing with a CO_2 laser is 1.5–3.5 times higher than when processing with a fiber optic laser, which is caused by more intense melt vibrations due to an increase in the penetration depth when processing with using a CO_2 laser.

The team of authors [62] used scanning electron microscopy, X-ray microanalysis, and optical metallography to study changes in the morphology, elemental composition, and microstructure of the surface layers of aluminium dodecaboride AlB12 subjected to high-power ion beam treatment. Under the influence of a superpower flow of particles, the surface layer of the irradiated material substantially changes its structural phase state and a new surface relief is formed. The topography of the surface of aluminium dodecaboride modified by a powerful pulsed beam of carbon ions is determined by the dose and density of the carbon ion flux power. The treated surface consists of two unequal areas irradiated by ion flows with different power densities. Most of the surface is irradiated with a more powerful stream and has a fused structure with numerous holes of various diameters, a smaller part consists of islands of tens of micrometers in size and contains precipitates of partially oxidized aluminium. It was established that the action of a powerful pulsed ion beam of an irradiated AlB12 surface leads to the formation of boron carbide.

V.S. Kovivchak et al. [63] analyzed the changes in the surface morphology of polycrystalline metals (magnesium, zinc and aluminium) upon repeated exposure to a powerful nanosecond ion beam with a current density of 50 to 150 A/cm². Thus, the effect of MIP on magnesium, zinc, and aluminium leads to the formation of regular structures, spheroidal ridges on ridges, and disk-shaped microparticles on their surface. The parameters of the observed morphological changes are determined by the ion beam current density, the number of irradiation pulses, and the type of material. The spatial parameters of the emerging relief are determined, which for the studied metals are in the range of 8–40 μm. The size of the particles formed during the formation of such a relief is from 0.1 to 1.5 μm. The appearance of regular structures is associated with excitation of capillary waves on the melt surface.

The method of laser hardening of the surface of parts [64], including heating the surface of the part with a laser beam using a scanner. The surface of the part is heated by a continuous laser while moving the beam along the normal to the vector of its movement with a beam oscillation frequency of 10 ÷ 200 Hz, oscillation amplitude $A = (2 ÷ 100)d$ and with a radiation energy density of 20 ÷ 26 W s /mm², where d is beam diameter on the surface of the part.

A known method [65] of laser hardening of the neck of the crankshaft is available. The method includes the relative movement of the surface of the workpiece and the laser source to ensure sequential projection of the laser spot on different parts of the processed surface area. The energy distribution of the effective laser spot is adapted in such a way that on a more heat-sensitive sub-section, such as a portion adjacent to the opening of the oil channel, it is different than on a less heat-sensitive sub-section to prevent overheating of the indicated more heat-sensitive sub-section.

There is a method [66] for obtaining a structured porous coating on titanium, which involves treating a titanium surface in an argon medium by moving a laser beam along it and simultaneously supplying titanium carbide powder to the irradiation zone, followed by acid etching by immersion in nitric acid for 3–5 days, washing and drying at a temperature of 50–100°C. In this invention, laser radiation with a power of not more than 300 W is used, the laser beam is moved on the surface at a speed of at least 20 mm/s, while using titanium carbide powder of a fraction of 80–100 μm

A method of producing [67] a multilayer modified surface of a titanium plate includes laser surface treatment of the plate sides, processing being carried out on both sides of the plate alternately with a 5 kW multi-channel diode laser. The reinforcing tracks in the form of a grid are applied by means of laser radiation paths along the same track, and inert gas is simultaneously supplied to the laser radiation affected area. The uniformity of the structure, hardness and depth of the hardened layer of the titanium plate is ensured.

A method [68] to obtain a coating of microstructured titanium carbide on the surface of titanium products is available. Titanium articles are placed in a reaction medium using a saturated hydrocarbon, and the surface of the article is treated with femtosecond laser radiation in the near infrared region of the spectrum with a pulsed power density of 10^{17} W/m^2 and 10% overlap of the laser exposure areas.

A method [69] of surface hardening of metals by changing the level of heat exposure on the treated surface, including local hardening by a scanning laser beam is available. The essence of the method is that the laser beam is polarized into a strip with a variable radiation intensity and scanned along this strip. The degree of polarization is set according to the accepted conditions of heat exposure, taking into account the fixed scanning speed, and the isothermal holding stages of the treated area at various temperature levels are successively implemented, and the temperature is changed at optimal speeds for the hardened metal. A method has been developed [70] for laser heat treatment of complex spatial surfaces of large-sized parts, including the action of a continuous laser beam focused in a light spot on the surface of the part and the application of parallel hardening tracks with overlapping by moving the light spot with a constant linear speed. The application of parallel hardening tracks on vertical or inclined surfaces is carried out by a beam directed to the workpiece at an angle of 0.5–5° from the perpendicular to the specified surface upward in the workpiece plane and when the process gas is supplied under pressure with a nozzle providing crystallization of the molten metal and balancing the gravity of a drop of melt. The application of parallel hardening tracks is carried out alternately in different hardening strips, spaced from each other at a distance sufficient for the tracks to cool at the set processing speeds. The application of parallel hardening tracks is carried out by a laser unit equipped with a coordinate laser head.

1.3. High-dose ion implantation

Ion implantation is a low-temperature process by which the components of one element are accelerated into the solid surface of the plate, thereby changing its physical, chemical or electrical properties. This method is used in the manufacture of semiconductor devices and in the decoration of metals, as well as in materials science studies. Ion beam technologies use monoenergetic and polyenergetic ion beams of various chemical elements. Ion energy, flux (beam current), fluence (integral flux) are the main parameters of the ion beam, which are determined by the goals and objectives of the processing. Usually, beams with ion energies of up to 100 keV and fluences of up to 10^{18} cm^{-2} are used to modify materials, and for surface alloying by ion mixing, with fluences of up to 10^{17} cm^{-2}.

Studies are ongoing on the effect of ion implantation by nitrogen ions on the corrosive behavior of AA7075 aluminium alloy. Implantation of nitrogen, at 2×10^{17} N $\cdot$ cm^{-2} and an energy of 50 keV, promotes the formation of an AlN layer that increases the corrosion resistance of the aluminium alloy, as evidenced by corrosion tests. Further studies showed that over time, the value of corrosion resistance is restored to its original value. Implanted and non-implanted alloys show similar electrochemical parameters after 20 h immersion in an electrolyte solution. However, microscopic examination shows less corrosion damage to implanted samples after 72 h of immersion. Thus, it can be argued that the long-term effect of corrosion resistance depends only on the heat treatment conditions of AA7075 alloys [71]. A scientific team led by O. Girk [72] is studying the physical and mechanical characteristics of aluminium alloy 2024 and titanium alloy Ti–6Al–4V after irradiation with fluxes of helium ions (He) and argon (Ar). Structural analysis by scanning electron microscopy showed that a cone-shaped structure forms on the surface of the materials under study, everywhere, except for the case of Ti–6Al–4V irradiation with He ions. The main factors contributing to the formation of such structures are relaxation of surface stresses and activated surface mobility of atoms. Elemental analysis shows that after irradiation with an ion beam does not change the chemical composition of the target. The study using a confocal microscope showed that the average roughness of the irradiated surface after irradiation with an Ar ion beam is higher than that of an He ion beam and higher on samples of alloy 2024 compared to Ti–6Al–4V alloy. The authors explain this result by the fact that the growth of surface structures is more related to the surface diffusion of atoms and not to their place in the atomic lattice as a result of collisions with a target [73].

The formation of chemical and phase composition, atomic structure, surface topography, mechanical and operational properties of surface layers of metal materials by ion implantation is one of the areas of modern science and technology [74]. Ion implantation allows you to increase the operational characteristics of the protected parts through surface alloying by an amount from several to tens of micrometers. Despite studies in this direction, the processes of formation of these layers, the structural mechanisms of their implementation, and the nature of the change in various properties of metals and alloys as a result of ion irradiation are still not clarified.

A widely used material in the aviation and rocket and space industry is titanium and alloys based on it, due to its unique physical and mechanical properties. The scientific group led by D.A. Alexandrov studied the hardening VT6 titanium alloy by ion implantation with nitrogen and ion saturation of the surface in the plasma of an aluminium alloy. The results of mechanical tests showed that the microhardness of the surface layers of the VT6 alloy increased by 1.5 times, compared with the starting material, and an increase in resistance to abrasive wear by more than 4 times was also observed. Structural analysis of the processed samples suggests that the hardening effect is achieved by changing the surface layer. The authors suggest that an increase in microhardness and a decrease in the lattice parameters of α-Ti and β-Ti are a consequence of solid solution hardening of the titanium base [75], which is indirectly confirmed by the correlation of the obtained experimental results with the results obtained in [76]. Thanks to modern research methods, namely atomic force microscopy, X-ray photoelectron spectroscopy, X-ray diffractometry, it was possible to study the surface profile, elemental and phase composition of VT6 titanium alloy subjected to ion-implantation of argon in a pulse-periodic mode and heat treatment. Initially, an aluminium film was deposited on the titanium alloy samples by magnetron sputtering, after which the surface was bombarded with argon ions. Investigation of the surface topology by atomic force microscopy and X-ray photoelectron studies of the elemental composition of thin surface layers showed that ion implantation of argon in a repetitively pulsed mode does not lead to a change in the surface morphology and mixing of the film with the main material. However, it is worth noting that further annealing with cooling in the furnace leads to the formation of strengthening phases, due to which the microhardness of the surface layers of the material increases by more than 30% [77].

Further work by a team of scientists with VT6 alloy was aimed at studying the effect of ion-beam mixing of carbon on the composition, atomic structure, and microhardness of surface layers. The authors argue that ion-beam mixing occurs at the substrate/film interface as a result of ion irradiation. A disordered carbon structure is formed, and conditions for the formation of titanium carbides are created in the transition region of the film/substrate system during ion beam treatment. The formation of a disordered structure of carbon and titanium carbides causes an increase in the microhardness of irradiated samples by a factor of 2–3 [78].

Today, of great interest is the ionic synthesis of interstitial phases in the surface layers of materials based on Ti and Al, in particular, aluminide phases (Ti3Al, TiAl, and TiAl3) [79]. The materials whose surface layers consist of aluminide phases with ultrafine crystallite sizes will be characterized by enhanced physical and mechanical properties. Technically pure titanium of the VT1-0 grade was used as a material for research in the work. As a result of the experimental part of the work, it was found that the starting material is an almost single-phase system. In an insignificant amount (0.9 vol%), β-Ti grains are present at the α-Ti grain boundaries, therefore, hardening by particles of the second phases can be neglected. Microstructural analysis of the irradiated samples shows that there is a decrease in both the longitudinal and transverse sizes of crystallites, and the magnitude of the change depends on the radiation dose, the higher the radiation dose, the smaller the grain size. Implantation of titanium with aluminium ions led to a change in the phase composition of the alloy. Lamellar precipitates of the Ti3Al phase with an ordered structure with an hcp crystal lattice were observed along the longitudinal grain boundaries of α-Ti. At the joints of α-Ti grains, precipitates of a TiAl3 phase with a bcc lattice are formed; inside the grains of α-Ti, small rounded precipitates of TiC carbide and TiO_2 oxide form on dislocations. Based on this, it can be concluded that implantation with aluminium ions led to the formation of a number of new phases, as a result of which the mechanical characteristics of the VT1-0 titanium alloy were increased [79].

There is a method [80] for hardening a nanostructured metal layer, which includes ion implantation with a dose of 10^{18} ion/cm^2 by polyenergy ions with an energy 15–60 keV, after ion implantation, the metal surface is ultrasonically treated at a frequency of 10 MHz, a strain amplitude of 10^{-5} and a processing time of 10^4 s. The invention relates to mechanical engineering and can be used to increase the strength characteristics of the material. The result: increase of operational characteristics and hardening of the metal, increase in the elastic modulus of the nanostructured metal layer.

A method has been developed [81] for the surface treatment of titanium alloy products. Two conductive materials are placed in the treatment zone – the first of titanium and the second of nickel-based alloy, the accumulation and diffusion on the surface of the product is carried out from the ions of the first conductive titanium material with a negative potential on the product of 120–200 V in a reaction gas medium consisting of a mixture of oxygen and argon in the ratio (1–2):1 at a pressure of 0.05–0.3 Pa, after which the supply of the reaction gas is stopped, an arc is excited on the second conductive material-alloy based on nickel and the ions of the second conducting material are accumulated with a negative potential on the product 15–20 V.

The method is available [82] of applying wear-resistant coatings to the blades of a gas turbine compressor by depositing alternating layers of metals and their nitrides with cleaning the surface of the blades with argon ions and ion implantation during the deposition process. The application of alternating coating layers is carried out using single-element cathodes of Ti, Al and/or composite cathodes – TiAl in a reaction gas – nitrogen. TiAlN layers are deposited with the following stoichiometric composition: Ti (23–28)%, Al (23–28)%, N (44–54)%, and with the structure of the TiN–Ti layer, (44–54)%, N (56–46)%. The composition of the multilayer coating is obtained at vacuum-arc discharge currents on single-element cathodes: on Al (75–83) A, on Ti (100–115) A. The coating thickness is applied on the order of 10 μm. The number and thickness of the layers included in the coating, set the speed of movement of products from one plasma source to another.

The method [83] of ion-plasma nitriding of titanium or titanium alloy products is available. Plasma generation is carried out in a frequency-pulsed glow discharge at a pulse repetition rate of 1 Hz - 100 kHz with a duty cycle fill factor of 10–90% at a pressure of 0.1-10 Pa, and the product is nitrided at an average ion current density of plasma 1–15 mA/cm^2 and at a pulsed ion current density of 5–100 mA/cm^2 in a heated state at a temperature that ensures the diffusion of nitrogen deep into the titanium or titanium-containing alloy. The injection of electrons into the working volume of the chamber is carried out in a pulsed mode, while the pulses of electron current are synchronized with the pulses of the glow discharge current. As the hollow cathode, the inner walls of the vacuum chamber or the surface of the heat shield are used.

1.4. New ways to increase the service characteristics of alloys (a combination of different methods of exposure)

As a result of a review of literary sources, it was found that the effect of external energy influences on the structure and properties of light alloys is a very urgent study task of modern materials science. However, the rapid development of scientific and technological progress establishes increased requirements for modern structural materials and single-stage processing is not enough. In this regard, more and more teams appear and, as a result, work combining various types of exposure with concentrated energy flows. Combined processing is understood as a combination of two or more methods of surface treatment of materials. The fastest growing area is coating, followed by laser treatment or processing with a high-intensity electron beam. Due to such two stages of processing, it is possible to dope the surface layers of the material by remelting and mixing the coating with the substrate. Through direct surface modifications using a laser and/or a high-intensity pulsed electron beam, nanocrystalline, amorphous structures and a supersaturated solid solution can be formed [84–86]. These modified layers have better characteristics in terms of physical and mechanical properties, which are often unattainable using conventional surface treatment methods. In addition, many researchers have succeeded in introducing pulsed melting methods on previously applied coatings or onlays of precious metals, which leads to highly concentrated alloying and improved surface properties. Numerous studies have shown that electron beam irradiation is an effective surface treatment method for film systems, including the Al (film) –Ti (substrate) system [87], etc. In [88], a study was made of the change in the corrosion resistance of the W–Al system coatings obtained by sputtering on the surface of an aluminium alloy with further processing by a high-intensity pulsed electron beam. Irradiation was carried out at room temperature with the following parameters: electron energy 27 keV, pulse duration 1.5 ms, beam energy density 4 J/cm^2, beam diameter 60 mm, vacuum 5.0×10^{-3} Pa, the number of pulses varied. The results showed that with an increase in the number of pulses, the number and density of surface defects decreases, tungsten particles, under the influence of beam energy, dissolve in an aluminium matrix, forming an alloying layer. After irradiating the surface with 35 pulses, the doping layer consisted mainly of ultrafine particles of Al4W and had a thickness of 10 mm. Corrosion tests showed that a corrosion-resistant coating forms on the surface of the material, this is due to the formation of a more stable oxide layer, a decrease in cathode activity and a modified microstructure. As it turned out, a combination of high-energy impact methods is possible not only when creating special coatings on the surface of the material, but also during the welding of aluminium alloys. Electron beam welding has a high energy density, a small amount of heat input during the welding process, high speed and significant penetrating power. It is usually carried out in a vacuum chamber, which eliminates the influence of air during the welding process and improves the quality of the joints. Investigations of welded joints obtained on 2A12 thick-sheet aluminium alloy showed that in the process of electron beam welding there is a profuse penetration of the metal with the formation of a pit of 2 mm, the weld structure was characterized by coarse grains, a combination of these factors reduces the strength characteristics of the final product. Further processing of the welded joint with an electron beam made it possible to increase tensile strength, in comparison with the initial alloy. The authors attribute this to the refinement of the microstructure of the weld. The average tensile strength of the joint increased by 31.2% to 383 MPa, and the fatigue strength also increased [89].

Along with aluminium, titanium and alloys based on it are widely used in industry, but titanium is also not without drawbacks, which leads to studies to improve its mechanical and physical characteristics. Spraying coatings using ceramic materials is a promising method for improving the surface characteristics of titanium [90]. The properties of such coatings strongly depend on their phase composition, grain size, porosity and its distribution. The use of post-treatment with high-energy energy sources (laser, electron beam, welding methods) seems to be a very interesting solution to eliminate the shortcomings of the formed coatings [91]. In [92], Al_2O_3–TiO_2-type coatings were deposited by thermal spraying onto the surface of a titanium alloy and were additionally irradiated with an electron beam. The experimental results showed that the deposition and electron-beam remelting of Al_2O_3–TiO_2 coatings on the titanium surface is an effective method of increasing the wear resistance of a titanium alloy without significantly reducing corrosion resistance. The microhardness of the remelted coating increased by about 50%. Thus, electron-beam processing can be used to obtain compact and uniform coatings with higher adhesion to the substrate.

When choosing structural materials, priority is given to materials based on intermetallic compounds. Intermetallic-based systems have unique properties, such as: low density, oxidation resistance, heat resistance, high specific strength and high melting point, which leads to their application in the automotive and aviation industries. In [93], the problem of creating protective coatings of the Ti–Al system on a substrate of technically pure VT1-0 grade titanium is considered. The coatings of the system under study were deposited on a titanium substrate by electron beam welding under various conditions. After coating, studies were carried out on the structure and mechanical properties of the samples. The research results showed that the microhardness of coatings is many times greater than the microhardness of a titanium substrate, and the wear resistance was also increased. Structural analysis showed that the electron beam current has a key influence on the structure and phase composition of the resulting coatings. Titanium and its alloys were not chosen as an object of research by chance, since it is of great importance for the aviation and aerospace industry. One of the areas of combined processing is a set of methods of plastic deformation of materials in order to improve the structure, and as a result, increase the mechanical characteristics of condensed matter.

In [94], experimental results of the effect of combined pressure treatment on the structure and properties of titanium VT1-0 were presented. At the first stage of processing, the samples were screw pressed, after which rolling was performed. It is shown that with this scheme of plastic deformation, there is not only an increase in the strength characteristics of the studied material, but also plasticity also increases. For the manufacture of elements of sheet structures for aviation purposes, it is necessary to use titanium alloys of medium and high strength [95]. The classic titanium alloys used for these purposes are VT5, VT20 and VT6 alloys. The combined treatment consisted of hydrogenation of the surface and the subsequent pressure treatment in the hydrogenated state, which was carried out by rolling at temperatures of the ($\alpha + \beta$) region. Experimental results showed that an increase in the amount of hydrogen in alloys from 0.3 to 0.7% (by mass) leads to the formation of a heterophase structure and a decrease in the size of the structural components of the α phase after deformation and vacuum annealing [96]. In addition to the aviation industry, VT6 titanium alloy has been widely used in medicine as an implant. But despite the active use of this alloy, there are unresolved problems in terms of hanging biocompatibility. Studies were conducted on the combined method of volumetric hardening and surface modification, including high-speed electrothermal treatment, vacuum aging, and the subsequent deposition of an oxide coating by the electrochemical method with the introduction of additional elements P and Ca. The processing results showed that this method of energy exposure allows you to get an alloy – tensile strength, which is more than 1200 MPa with a fairly high plastic characteristics. The resulting coating on VT6 titanium alloy is biocompatible and toxicologically safe, which generally suggests that this material processing technology can be used to create high-quality implants [97].

A method was developed [98] for producing a wear-resistant and thermodynamically stable multilayer coating based on refractory metals and their compounds, including vacuum-plasma deposition of coating layers based on refractory metals of titanium, zirconium and their compounds on a previously cleaned surface of the substrate. All layers of the coating are applied by magnetron sputtering, the first adhesive layer of titanium being applied by magnetron sputtering of a titanium target in an inert gas medium. The next layer of titanium nitride TiN is applied by sputtering a titanium target in a gas mixture of inert and reaction gases, then alternating layers of two-component zirconium nitride ZrN are deposited by sputtering a zirconium target in a gas mixture of inert and reaction gases and zirconium by sputtering a zirconium target in an inert gas. Then alternating layers of ternary titanium and zirconium nitride TiZrN are applied by simultaneously sputtering titanium and zirconium targets in a gas mixture of inert and reaction gases and zirconium by sputtering a zirconium target in an inert gas. Alternating layers of two-component zirconium nitride ZrN and zirconium are applied in the following sequence: a zirconium layer, a zirconium nitride layer ZrN, a zirconium layer, a zirconium nitride layer ZrN, a zirconium layer. Alternating layers of ternary titanium and zirconium nitride TiZrN and zirconium are deposited in the following sequence: a layer of titanium nitride and zirconium TiZrN, a layer of zirconium, a layer of titanium and zirconium TiZrN, a layer of zirconium, a layer of titanium nitride and zirconium TiZrN.

The method [99] is used to produce a wear-resistant multilayer coating, including ion cleaning with heating and thermomechanical activation of the substrate using an electric arc evaporator in nitrogen by its ion bombardment with an energy of 0.8–1.0 keV before deposition and vacuum-plasma deposition in nitrogen multilayer coating. In this case, the deposition of alternating layers is repeated at least three times. The surface of the substrate is cleaned with nitrogen ions in a glow discharge with non-contact heating of the surface by a resistive heater up to 400–430 K for 30 min, then, during ion cleaning, thermomechanical activation and heating of the surface of the substrate with titanium ions to 665–695 K.

Extensive work has been carried out by Russian scientific teams on the study of the physicomechanical properties of titanium and its alloys after various surface and volume impacts. Initially, a lower layer of titanium nitride TiN with a polycrystalline structure is applied for 3 min with a final coating temperature after deposition of 670–700 K, then alternating layers of two-component titanium nitride TiN with a nanocrystalline structure and three-component titanium nitride and aluminium Ti–Al–N with a nanocrystalline structure are, and the TiN layer with a nanocrystalline structure is applied by evaporation of two titanium cathodes when the coating is heated to a temperature of 680–710 K for 3 min, and the Ti–Al–N layer with a nanocrystalline structure is applied by simultaneous evaporation of two titanium and one aluminium cathode when the coating is heated to a temperature of 690–720 K during the same time, while alternating layers are deposited until the temperature of the upper layer reaches 730–760 K, with the Ti–Al–N layer being applied last.

A method [100] is available for producing a multilayer coating, comprising ion cleaning with heating the substrate before deposition and vacuum-plasma deposition of a multilayer coating. The substrate is heated before deposition of the coating in the process of ion cleaning with a heating rate of 70 K/min to a temperature of 650 K, a lower layer of Ti titanium is applied to the substrate in an inert gas medium with a final coating temperature after deposition of 615÷630 K, then in an argon gas mixture and alternating layers of bicomponent titanium nitride TiN with a polycrystalline structure and bicomponent titanium nitride TiN with a nanocrystalline structure, and a TiN layer with a polycrystalline the structure is applied with a gradual increase in the temperature of the coating to 645–675 K with a heating rate of 3.2–4.4 K/min, and a TiN layer with a nanocrystalline structure is applied with a gradual increase in the temperature of the coating to 695–725 K with a heating rate of 3.7 K/min, while the deposition of coating layers is carried out by the evaporation of two titanium cathodes and the deposition of alternating layers is repeated no more than two times, with the latter being applied a layer of bicomponent titanium nitride TiN with a nanocrystalline structure.

The invention [101] relates to the field of mechanical engineering, and in particular to methods for modifying the surface treatment of articles made of titanium alloys to improve their tribological characteristics. The method includes preliminary cleaning and activation of the surface of a titanium alloy product by bombarding it with argon ions using a gas plasma generator and ion-plasma deposition of the composite coating by magnetron sputtering of a cathode containing titanium carbide and molybdenum disulphide, with the application of a negative potential to the product and combining the deposition process with bombardment of the surface by argon ions. The coating is deposited on a pre-alloyed surface layer of a titanium alloy, which is formed by magnetron sputtering of the cathode material, while the surface is bombarded with argon ions using a gas plasma generator and a negative potential is applied to the product that exceeds the potential value during coating deposition. The initial ratio of the components of the sprayed cathode material is:

titanium carbide 40–60 wt.%, molybdenum disulphide – the rest. The result: reduced coefficient of friction and increased wear resistance.

A method was developed [102] for producing a multilayer composite nanostructured coating, including placing the substrate in a vacuum chamber, ion etching of the substrate, and deposition of the material onto the substrate by PVD in a working gas medium. At least two electric arc plasma sources with flow separation are used for material deposition, one of them is equipped with a cathode of refractory metal. Ion etching is carried out in an inert gas medium by the aforementioned electric arc plasma source with a refractory cathode with flow separation during the formation of a pulsed gas discharge. Before ion etching, a metal layer is deposited on the substrate with a thickness greater than the size of microroughnesses on the surface, and ion etching is carried out before etching the deposited layer. Ion etching is carried out in a mixture of an inert and chemically active gas by the aforementioned electric arc plasma source with a refractory cathode with flow separation during the formation of a pulsed gas discharge. When coating, the substrate for each cycle is under each plasma source for a time not exceeding $t100V$ s, where V is the deposition rate of the coating, nm/s. When applying the coating, the time interval t_p between pulses of the aforementioned pulsed gas discharge satisfies the condition $t_p \leq \delta_0/C$, where δ_0 is the thickness of the monatomic coating layer; C is the coating rate, and the pulse duration is $\tau_p \geq k \cdot t_p$, where $k = \varepsilon/v \cdot e$, ε is the atomic displacement energy in the crystal lattice of the coating material; v is the pulse amplitude, and e is the electron charge. The parameters of the pulses of the mentioned pulsed gas discharge satisfy the condition $\tau_p < t_p$, where τ_p is the pulse duration, and t_p is the time interval between pulses.

There is a method [103] of applying an ion-plasma coating to an instrument, comprising placing the workpiece in a vacuum chamber from which air is pumped out to operating pressure, then ion cleaning, heating the surface of the workpiece and applying a multilayer wear-resistant coating. Activation and ionic cleaning of the surface of the processed tool is carried out by a high-current plasma source with a filament cathode and electric arc evaporators in an inert argon gas medium when the surface is heated to a temperature of 300-450°C. The multilayer wear-resistant coating consists of a lower titanium layer and a Ti–Al intermetallic-based coating layer, which is applied by two electric arc evaporators with titanium and aluminium cathodes when the process is assisted by a high-current plasma source with a filament cathode in a nitrogen atmosphere.

The coating based on the intermetallic Ti–Al system is carried out in one technological cycle.

The method [104] of hardening the surface of products made of metal materials includes processing with a stream of high-temperature pulsed self-stabilizing plasma. Each pulse action of the plasma flow on the surface of the product is accompanied by the action of an electric current from a capacitively inductive storage device that commutes the plasma flow between the plasma generator electrode and the hardened surface, while controlling the change in the resistance and speed of the commuting jet of the plasma flow with providing a pulse connection of the hardened surface as a cathode or anode with the optimal number and sequence of pulses of impact on the hardened surface of the product of the plasma flow and with simultaneous action of acoustic vibrations on the surface and the reinforcing metal vapour introduction into the plasma stream. The resistance and speed of the commuting jet of the plasma flow is changed by controlled pulse input into the interelectrode gap of the plasma generator of a mixture of combustible gas with air. Propane butane is used as combustible gas. The hardened surface is affected by acoustic vibrations that have a sound pressure level in the range of 140–150 dB and a frequency of up to 20 000 Hz. Metal vapours are introduced into the plasma from the erodible metal electrode when it is connected as an anode. A suspension containing alloying elements is introduced into the interelectrode gap of the plasma generator.

Having reviewed the current state of research in the field of modifying the structure and properties of light metallic materials, it can be stated that this area of research is relevant and requires further development. The importance of improving the performance characteristics of aluminium-based alloys is also dictated by the requirements of practice, since gradually light metals displace steel in the manufacture of parts and assemblies for various purposes.

References for Introduction and Chapter 1

1. Arsenault R. J., Treatise on Materials Science and Technology, Plastic Deformation of Materials. University of Maryland, College Park, MD, USA, 1975.
2. Hao Y., Gao B., Tu G.F., Cao H., Hao S.Z., Dong C., Applied Surface Science. 2012. V.258. p. 2052-2056.
3. Wong T.T., Liang G.Y., Journal of Materials Processing Technology. 1997. V. 63. p. 930-934.
4. Sharkeev Yu.P., Legostaeva E.V., Panin S.V., Gritsenko B.P., Surface and Coatings Technology. 2002. V.158-159. p. 674-679.
5. Ehtemam-Haghighi S., Attar H., Dargusch M.S., Kent D., Journal of Alloys and Compounds. 2019.V.787. p. 570-577.
6. Lipinski T., Manufacturing Technology. 2015. V.15. No.4. M2015100.
7. 7. Władysiak R., Kozuń A., Archives of Foundry Engineering. 2015. V.15. Issue 1. p. 113-118.
8. Ghyngazov S.A., Vasil'ev I.P., Surzhikov A.P., Frangulyan T.S., Chernyavskii A.V., Technical Physics. 2015.V.60. No.1. p. 128-132.
9. Yang Y., Frazer D., Balooch M., Hosemann P., Journal of Nuclear Materials. 2018.V.512. p. 137-143.
10. Todaro C.J., Easton M.A., Qiu D., Wang G., Stjohn D.H., Qian M., Journal of Materials Processing Tech. Journal of Materials Processing Tech. 2019.V. 271 p. 346-356.
11. Nedyalkov N., Stankova N.E., Koleva M.E., Nikov R., Aleksandrov L., Iordanova R., Atanasova G., Iordanova E., Yankov G., Applied Surface Science. 2019.V.475 p. 479-486.
12. Kuznetsov G.V., Feoktistov D.V., Orlova E.G., Batishcheva K., Ilenok S.S., Applied Surface Science. 2019.V.469 p. 974-982.
13. Mori Y., Sunahara A., Nishimura Y., Hioki T., Azuma H., Motohiro T., Kitagawa Y., Ishii K., Hanayama R., Komeda O., Sekine T., Takeuchi Y., Kurita T., Miura E., Sentoku Y., Journal of Physics D: Applied Physics. 2019.V.52. 105202.
14. Huang Y., Yao Z., He C., Zhu L., Zhang L., Bai J., Xu X. Journal of Physics: Condensed Matter. 2019.V.31. No.15. 153001.
15. Budovskikh E.A., Gromov V.E., Romanov D.A., Doklady Physics. 2013. V. 58. No.3. p. 82-84.
16. Romanov D.A., Budovskikh E.A., Gromov V.E., Journal of Surface Investigation. X-ray, Synchrotron and Neutron Techniques. 2011. V.5. No. 6. p. 1112-1117.
17. Rotshtein V., Materials surface processing by directed energy techniques. Ed. by Pauleau Y. Oxford: Elsevier, 2006. Ch. 6.
18. Gromov V.E., Ivanov Yu. F., Vorobiev S. V., Konovalov S. V., Fatigue of steels modified by high intensity electron beams. Cambridge International Science Publishing, 2015.
19. Engelko V., Yatsenko B., Mueller G., Bluhm Y., Vacuum. 2001. V.62. p. 211-216.
20. Ozur G.E., Proskurovsky D.I., Rotshtein V.P., Markov A.B., Laser and Particle Beams. 2003.No.21. p. 157-174.
21. Rotshtein V.P., Ivanov Yu.F., Proskurovsky D.I., Karlik K.V., Shulepov I.A., Markov A.B., Surface and Coatings Technology. 2004. V. 180-181. p. 382-386.
22. Ivanov Yu.F., Rotshtein V.P., Proskurovsky D.I., Orlov P.V., Polestchenko K.N., Ozur G.E., Goncharenko I.M., Surface and Coatings Technology. 2000. V.125. No.1-3. p. 251-256.
23. Proskurovsky D.I., Rotshtein V.P., Ozur G.E., Ivanov Yu.F., Markov A.B., Surface

and Coatings Technology. 2000. V.125. p. 49-56.

24. Utu D., Marginean G., Colloids and surfaces a: physicochemical and engineering aspects. 2017. V.526. p. 70-75.

25. Tang G., Luo D., Tang S., Mu Q., Wang L., Ma X., Journal of Alloys and Compounds. 2017. V.714. p. 96-103.

26. Jung A., Buchwalder A., Hegelmann E., Hengst P., Zenker R., Surface and Coatings Technology. 2018.V.335. p. 166-172.

27. Ding S., Xue M., Wu R., Lai Y., Men Y., Kong X., Han Li, Han J., Yang W., Yang Y., Du H., Wang C., Yang J., Journal of Alloys and Compounds. 2018.V.744. p. 615-620.

28. Chesov Yu.S., Zverev EA, Plohov AV, Metal processing (technology, equipment, tools). 2010. No. 1. C. 14-18.

29. Chyosov Yu.S. Zverev EA, Tregubchak PV, Metal processing (technology, equipment, tools). 2012. No. 1. S. 10-13.

30. Li Y., Zhang P., Bai P., Wu L., Liu B., Zhao Z., Surface and Coatings Technology. 2018.V. 334. p. 142-149.

31. Wan Q., Rui X., Wang Q., Bai Y., Du Z., Niu W., Wang W., Wang K., Gao Y., Surface and Coatings Technology. 2019.V. 367. p. 288-301.

32. Ardila-Rodríguez L.A., Menezes B.R.C., Pereira L.A., Takahashi R.J., Oliveira A.C., Travessa D.N., Surface and Coatings Technology. 2019.V. 377. 124930.

33. Kubatík T.F., Lukáč F., Stoulil J., Ctibor P., Průša F., Stehlíková K., Surface and Coatings Technology. 2017. V. 319. p. 145-154.

34. Koshuro V. A., Fomin A. A., Rodionov I. V., Fomina M. A., Physical chemistry of the surface and protection of materials. 2018.V. 54. No. 5. C. 499-504.

35. Zhuravina T.V., Bataev I.A., Ruktuev A.A., Alkhimov A.P., Lenivtseva O.G., Butylenkova O.A., Metal processing. 2012.V. 54. No. 1. S. 90-95.

36. 36. Golkovsky M.G., Samoilenko V.V., Popelyukh A.I., Ruktuev A.A., Plotnikova N.V., Belousova N.S., Metal processing. 2013.V. 61. No. 4. P. 44-48.

37. Roshchin M.N., JARiTS. 2019.Vol. 1. No. 14. S. 27-30.

38. Mikheev A.E., Girn A.V., Ivasev S.S., Vakheev E.V., Procedure for strengthening surface of items of titanium alloys. Patent RF, No. 2427666, 2011.

39. Vinogradova T. S., Tarakanova T. A., Farmakovskij B. V., Ulin I. V., Sholkin S. E., Jurkov M. A., Method of making catalytic composite coating. Patent RF, No. 2417841, 2011.

40. Khusnimardanov R. N., Vardanyan E. L., Nazarov A. Y., Ramazanov K. N., Bryukhanov E. A., Method of hardening of cutting tool by deposition of multilayer coatings of system Ti-Al. Patent RF, No. 2700344, 2019.

41. Antsiferov V.N., Kameneva A.L., Method of obtaining complex nitride-based coating. Patent RF, No. 2429311, 2011.

42. Romanov D.A., Olesjuk O.V., Budovskikh E.A., Gromov V.E., Method of electro-explosive sputtering of composite wear-resistant coatings of system TiC-Mo on friction surface. Patent RF, No. 2518037, 2014.

43. Kovtunov A.I., Bochkarev A.G., Gushchin A.A., Khokhlov Y. Y., Method of facing of titanium and titanium alloys of heat-resistant and wearresistant coatings based on titanium aluminides. Patent RF, No.2699474, 2019.

44. Kovtunov A.I., Ostryanko A.M., Bochkarev A.G., Method of intermetallid alloys facing on the basis of the titan-copper system. Patent RF, 2670317, 2018.

45. Reboul M.C., Baroux B., Materials and Corrosion. 2010. V. 62. Issue 3. p. 215-233.

46. Principe E. L., Shaw B. A., Davis G. D., Corrosion science. 2003. V. 59. Issue 4. p. 295-313.

47. Elahi M. R., Rajamure R. S., Vora H. D., Srinivasan S. G., Dahotre N. B., Applied Surface Science. 2015. V. 328. p. 205-214.

48. Xia H., Zhang C., Lv P., Cai J., Jin Y., Guan Q., Nuclear Instruments and Methods in Physics Research Section B: Beam Interactions with Materials and Atoms. 2018.V. 416. p. 9-15.

49. Yu P., Yan M., Tomus D., Brice C. A., Bettles C. J., Muddle B., Qian M., Materials Characterization. 2018.V. 14.p. 43-49.

50. Wang J.T., Xie L., Luo K.Y., Tan W.S., Cheng L., Chen J.F., Lu Y.L., Li X.P., Ge M.Z., Surface and Coatings Technology. 2018.V. 349. p. 725-735.

51. Gao Y.K., Materials Science and Engineering: A. 2011. V. 528. Issue 10-11. p. 3823-3828.

52. Liu P., Sun S., Hu J. Optics and Laser Technology. 2019.V. 112.p. 1-7.

53. Li J., Zhou J., Sun Y., Feng A., Meng X., Huang S., Sun Y., Optics and Laser Technology. 2019.V. 120.105670.

54. Gottardi G., Tocci M., Montesano L., Pola A., Wear. 2018.V. 394-395. p. 1-10.

55. Tong Z., Jiao J., Zhou W., Yang Y., Chen L., Liu H., Sun Y., Ren X., Surface and Coatings Technology. 2019.V. 377. 124799.

56. Wang J., Lu Y., Zhou D., Sun L., Xie L., Wang J., Vacuum. 2019.V. 165.p. 193-198.

57. Zhirkova ON, Morozov AP, Basic research. 2006. No. 6. S. 100.

58. Zhirkova ON, Morozov AP, Basic research. 2006. No. 5. P. 62.

59. Muratov VS, Morozova EA, International Journal of Applied and Basic Research. 2011. No.5. S. 41.

60. Proskuryakov V. I., Rodionov I. V., Novikov M. V., Progressive technologies and processes. 2019.S. 224-228.

61. Savrai R. A., Malygina I. Yu., Makarov A. V., Osintseva A. L., Rogovaya S. A., Davydova N. A., Diagnostics, Resource and Mechanics of materials and structures. 2018. No. 5. P. 86–105.

62. Potemkin GV, Lepakova OK, Physics and chemistry of material processing. 2014. No. 6. P.5-12.

63. Kovivchak V.S., Panova T.V., Mikhailov K.A., Proceedings of higher educational institutions. Powder metallurgy and functional coatings. 2012. No1. S. 48-51.

64. Biryukov V.P., Gudushauri E.G., Tatarkin D.Y., Fishkov A.A., Method of laser hardening of parts surface. Patent RF, No.2684176, 2019.

65. Gabilondo A., Dominguez J., Soriano C., Ocana J. L., Method and system for surface laser strengthening of the processed item. Patent RF, No.2661131, 2018.

66. Zhevtun I. G., Yarusova S. B., Gordienko P. S., Subbotin E. P., Method for obtaining structured porous coating on titanium. Patent RF, No.2669257, 2018.

67. Evstyunin G.A., Method of producing multilayer modified surface of titanium. Patent RF, No.2686973, 2019.

68. Abramov D.V., Arakeljan S. M., Kochuev D.A., Prokoshev V.G., KHorkov K.S., Method of micro-structured titanium carbide coating producing on surface of article from titanium or titanium alloy using laser radiation. Patent RF, No.2603751, 2016.

69. Ruzanov F. I., Pyrikov P. G., Method of surface hardening of metals. Pat. RF, No. 2004129402, 2006.

70. Sirotkin O.S., Blinkov V.V., Vajnshtejn I.V., Kondratjuk D.I., Chizhikov S.N., Kozhurin M. V., Method of later thermal processing of complex surfaces of large-sized parts. Patent RF, No.2425894, 2011.

71. Abreu C.M., Cristóbal M.J., Figueroa R., Nóvoa X.R., Pena G., Nuclear Instruments and Methods in Physics Research Section B: Beam Interactions with Materials and Atoms. 2019.V. 442. p. 1-12.

72. Bizyukov I., Girka O., Kaczmarek Ł., Klich M., Myroshnyk M., Januszewicz B., Owczarek S., Nuclear Instruments and Methods in Physics Research Section B: Beam Interactions with Materials and Atoms. 2018.V. 436. p. 272-277.

73. Girka O., Bizyukov O., Balkova Y., Myroshnyk M., Bizyukov I., Bogatyrenko S., Nuclear Instruments and Methods in Physics Research Section B: Beam Interactions with Materials and Atoms. 2018.V. 420.p. 49-46.

74. Kalin B. A., Volkov N. V., Oleinikov I. V., Bulletin of the Russian Academy of Sciences. The series is physical. 2012.V. 76. No. 6. S. 771-776.

75. Aleksanrov D. A., Muboyadzhyan S. A., Lutsenko A. N., Zhuravleva P. L., Aviation materials and technologies. 2018.V. 51. No. 2. S. 33-39.

76. Sharkeev Yu.P., Ryabchikov A.I., Kozlov E.V. et al., News of higher educational institutions. Physics. 2004.V. 47. No. 9. S. 44-52.

77. Vorobyov V.L., Bykov P.V., Bystrov S.G., Kolotov A.A., Bayankin V.Ya., Kobziev V.F., Makhneva T.M., Chemical Physics and Mesoscopy. 2013. V. 15. No. 4. S. 576-581.

78. Sozonova N.M., Vorobyov V.L., Drozdov A.Yu., Bayankin V.Ya., Chemical physics and mesoscopy. 2018.V. 20. No. 4. S. 569-575.

79. Kurzina I.A., Kozlov E.V., Sharkeev Yu.P. et al., Nanocrystalline intermetallic and nitride structures formed upon ion-plasma exposure. Tomsk: NTL Publishing House. 2008.

80. Artemov I.I., Akimov D.A., Krevchik V.D., Method of hardening a nanostructured metal layer by ion implantation with ultrasonic action. Patent RF, No.2699880, 2019.

81. Mubojadzhjan S.A., Nochovnaja N.A., Aleksandrov D.A., Gorlov D.S., Method of processing surface of article made from titanium alloy. Patent RF, No.2445406, 2012.

82. Gejkin V.A., Belova L.N., Nagovitsyn E.M., Poklad V.A., Sharonova N.I., Rjabchikov A.I., Stepanov I.B., Procedure for application of wear resistant coating on blades of compressor of gas turbine engine (GTE). Patent RF, No.2430992, 2011.

83. Denisov V.V., Koval N. N., Shchanin P.M., Ostroverkhov E.V., Denisova Y.A., Ivanov Yu.F., Akhmadeev Y.K., Lopatin I.V., Method of ion-plasma nitriding of articles from titanium or titanium alloy. Patent RF, No.2686975, 2019.

84. Cai J., Yang S.Z., Ji L., Guan Q.F., Wang Z.P., Han Z. Y., Surface and Coatings Technology. 2014. V. 251. p. 217-225.

85. Zhang C. L., Cai J., Lv P., Zhang Y. W., Xia H., Guan Q. F., Journal of Alloys and Compounds. 2017. V. 697. p. 96-103

86. Lv P., Sun X., Cai J., Zhang C. L., Liu X. L., Guan Q. F., Surface and Coatings Technology. 2017. V. 309. p. 401-409.

87. Otten C., Reisgen U., Schmachtenberg M., Welding in the World. 2016. V. 60. Issue 1. p. 21-31.

88. Zhang C., Lv P., Cai J., Zhang Y., Xia H., Guan Q., Journal of Alloys and Compounds. 2017. V. 723. p 258-265.

89. Chen G., Liu J., Shu X., Gu H., Zhang B., Feng J., Materials Science and Engineering: A. 2019. V. 744. p. 583-592.

90. Utu I.D., Marginean G., Hulka I., Serban V.A., Cristea D., International Journal of Refractory Metals and Hard Materials. 2015. V. 51. p. 118-123.

91. Dudek A., Iwaszko J., Archives of Materials Science and Engineering. 2008. V. 33. Issue 1. p. 39-44.

92. Utu I.D., Marginean G., Colloids and Surfaces A: Physicochemical and Engineering Aspects. 2017. V. 526. p. 70-75.

93. Matz O.E., Munkueva D.D., Bataev I.A., Actual problems in mechanical engineer-

ing. 2016. No3. S. 440-445.

94. Stolyarov VV, Salimgareev H.Sh., Soshnikova EP, Beigelsimer Y.E., Orlov D.V., Synkov S.G., Reshetov A.Yu., High-pressure physics and technology. 2003. T. 13. No. 1. S. 54-59.

95. Kablov E.N., Wings of the Motherland. 2016. No5. S. 8-18.

96. Nocturnal N.A., Panin P.V., Transactions of VIAM. 2017.V. 57. No. 9. S. 3-11.

97. Gordienko AI, Smyaglikov IP, Nazarova OI, Modern methods and technologies for the creation and processing of materials. 2017.S. 199-205.

98. Antsiferov V.N., Kameneva A.L., Method for obtaining wear-resistant and thermo-dynamically resistant multi-layer coating on the basis of high-melting metals and their compounds. Patent RF, No.2433209, 2011.

99. Kameneva A.L., Method of making sandwich wear-resistant coatings. Patent RF, No.2494170, 2013.

100. Kameneva A.L., Method of making sandwich coatings. Patent RF, No.2487189, 2013.

101. Savostikov V.M., Tabachenko A.N., Potekaev A.I., Dudarev E.F., Application of an-tifriction wear-proof coat on titanium, alloys. Patent RF, No.2502828, 2013.

102. Bashkov V.M., Beljaeva A.O., Dodonov A.I., Method and device to produce mul-tilayer composite nanostructured coatings and materials. Patent RF, No.2463382, 2012.

103. Ramazanov K.N., Vardanyan E.L., Nazarov A.Y., Bryukhanov E.A., Method of ap-plying wear-resistant coating by ion-plasma method. Patent RF, No.2694857, 2019.

104. Vasilik N.J., Kolisnichenko O.V., Tjurin J.N., Method of hard facing of metal items. Patent RF, No.2541325, 2015.

Research materials, experimental procedures, description of equipment and scientific approaches

2.1. Justification for the use of materials

It is known that aluminium is the most common of non-ferrous metals and takes first place in the content in the earth's crust, 81.3 kg of aluminium per 1 ton of the earth's crust, which is 31.3 kg more than iron [1–2]. In view of its prevalence, it causes increased interest among the scientific community in the direction of studying its properties and the possibilities of their modification [3–9]. One of the common methods of aluminium modification is the addition of alloying elements, for example, silicon, during casting [6–9].

Alloys of aluminium with silicon are comparable in strength and corrosion properties to stainless steel, but they are lighter than it, which, undoubtedly, is a plus. The strength properties of the alloy are provided by silicon, which is part of it, and corrosion resistance is guaranteed by the formation of a protective oxide film that occurs on the surface in oxidizing environments, one of which is oxygen. The alloy also wins in plasticity; it easily repeats the most complex forms, filling them evenly. As a result, casting of the alloy is simplified, which means that the production process is cheaper [6, 7, 10].

Eutectic alloys contain from 10 to 13% silicon, have moderate strength properties, but elongation quite high for cast alloys and are the most common cast alloys based on aluminium. The main advantage over other casting aluminium alloys is their very good casting properties, primarily high fluidity. These casting alloys are very well suited for casting thin-walled, complex in shape, sealed, resistant to vibration and shock loads of products [8–11].

In view of the above properties, the considered alloys are widely used as structural materials. In particular, the pistons of internal combustion engines and compressors are mainly made of eutectic and hypereutectic Al–Si alloys [7–12].

The constant race to increase the efficiency and environmental friendliness of modern engines forces designers to modify the connecting rod and piston group, in particular to reduce its weight, to change the geometric dimensions, thereby giving the pistons a more complex shape. At the same time, the loads on the connecting rod and piston group do not decrease due to the conservation of engine power, which leads to a significant reduction in resource. In this regard, the search for new effective methods for improving the properties of Al–Si alloys based on modification, microalloying, and surface treatment remains relevant [13–21].

In this work, we used an Al–Si alloy with a percentage of Al elements of 84.88% and 11.10% Si, established by the results of X-ray spectral analysis. The distribution of chemical elements in the initial alloy structure is shown in Fig. 2.1 *a*. The alloy of aluminium with silicon under consideration is widely used in the automotive industry for the production of pistons of internal combustion engines.

The samples for research had sizes $20 \times 20 \times 10$ mm^3 (Fig. 2.1 *b*) and were oriented perpendicular to external energy influences.

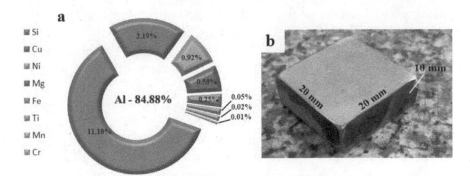

Fig. 2.1. Results of X-ray spectral analysis of Al–11Si–2Cu alloy (*a*), geometrical dimensions of sample (*b*).

2.2. Methods for determining the mechanical and physical properties of the investigated materials

As a characteristic of the mechanical properties of surface layers, we used one of the most accurate and sensitive methods – microhardness measurement. Its differences before and after processing can serve as an indicator of hardening of modified surface layers of metals and alloys. Microhardness measurements were carried out using an HVS-1000 microhardness meter and a Shimadzu DUH-211S ultramicrohardness meter using the micro-Vickers method [22].

The microhardness was measured in accordance with the requirements of International Standard ISO 6507:2018 (E) using the reconstructed print method (main) and a tetrahedral pyramid with a square base.

To measure microhardness, a metallographic thin section pre-etched to reveal the structure was examined under a microscope, a place for research was selected, a diamond indenter was applied to this place, a load was applied, after which the load was removed, the thin section was returned to the field of view of the microscope objective and the print was measured.

The load was constant and amounted to 0.5 N. Microhardness measurements were carried out in grain and eutectic. The time of applying and holding the load was 10 s, and the removal of the test load was 5 s.

The microhardness was measured both directly from the side of the subjected modification, and at different distances from it using cut-thin section. Using the method of measuring microhardness, you can quickly and accurately control the quality of products and materials, as well as carry out numerous physico-chemical studies related to the recognition of substances and the study of their properties, functions and structural transformations. This method receives a number of important applications in connection with the possibility of indirectly evaluating other mechanical characteristics of substances, between which there is a certain correlation between microhardness. The main difference between the microindentation method and hardness is the use of insignificant loads. The microhardness method is well combined with the study of the microscopic structure of the material.

Nanohardness was measured using a Super Nanoscan scanning nanohardness meter by sclerometry, which consisted of scratching the surface of the material and then scanning the image of the resulting prints. Previously, the shape of the tip of the NanoScan was calibrated on a reference material by applying a series of scratches at different loads. The value of the hardness of the material is calculated relative to the hardness of the standard by the ratio of the loads and the widths of the resulting scratches on the test and reference materials. Allows you to measure the hardness of materials on a submicron scale and the nature of their destruction.

The next stage of the study of the microstructure was carried out using an Olympus GX-51 optical microscope. In order to study the material using metallography, the samples were prepared as follows: the sample was cut, polished, polished, and etched. To create optical contrast, the samples were chemically etched with a solution containing 72% H_2O, 21% HF, and 7% HCl.

The tribological properties of the modified alloy were characterized by a wear coefficient and a friction coefficient (TRIBOtester devices). The installation is based on the principle of a tester, on a pin disk with a rotating disk and a static sample. The main element of the device is a pin having a certain radius and made of a certain material, which is aligned perpendicular to the rotating disk. By changing the load on the pin, it is possible to change the coefficient of friction and, therefore, determine what wear occurs between the materials (wear coefficient). The coefficient of friction is measured as the inertial moment arising between the investigated materials.

The load on the pin is provided by a weight that can be moved along the extended bracket. This provides an optimal form of load, and also makes it possible to change the load during testing.

2.3. Methods of analysis of changes in the fine structure and phase composition of surface layers

The study of the porosity of the coating and the heat-affected zone of the samples of the studied alloy was carried out using an NT-MDT Solver NEXT atomic force microscope. This method is one of the most powerful modern methods for studying the morphology and local properties of a solid surface with high spatial resolution [23], and one of its most popular varieties is atomic force microscopy (AFM) [24].

Using AFM, images were taken that measured the width of the surface layers, the shape and size of the grains were studied, and the conclusion was drawn that the coating was uniform.

Obtaining AFM images involves a specially organized process of scanning a sample. When scanning, the probe first moves over the sample along a certain line (line scan), then the probe returns to its original point and moves to the next scan line (frame scan), and the process repeats again. The feedback signal recorded in such a way during scanning is processed by a computer, and then the AFM image of the surface relief is constructed using computer graphics.

In an atomic force microscope, a study of the surface topography and its local properties is carried out using probes (cantilevers), which are made by liquid-phase etching.

The elemental and phase composition and the defective substructure of the modified layer were analyzed by scanning electron microscopy (SEM) using Philips SEM-515 equipped with an EDAX ECON IV microanalyzer and an analytical scanning electron microscope with a large camera and an enlarged motorized table TESCAN VEGA SB [25].

The TESCAN Vega SB scanning electron microscope has a tungsten cathode with thermionic emission, four Wide Field Optics lenses using an intermediate lens to optimize beam shape and size. The resolution in high vacuum mode was 3.0 nm at a voltage of 30 kV.

The photographs presented in this work were made in secondary electrons, in the mode closest to the optical image.

The chemical composition was determined using an INCAx-act energy dispersive detector for X-ray microanalysis. Elemental analysis of the individual phases was carried out by the method of electron probe microanalysis, which allows you to study the presence, content and distribution of elements of the periodic table.

Main characteristics of the Philips SEM-515 microscope: wide range of accelerating voltage of 3–30 kV and beam current; resolution up to 10 nm (at accelerating 30 kV); the minimum increase is ~ 20 times, the maximum is 160 000 times; depth of field, corresponding to the maximum resolution of the final image elements (0.2 mm) by the human eye, is about 0.5 of its linear dimensions; analyzed elements – starting from fluorine; detection limit 0.2% weight. (depends on the set of analyzed elements); extreme accuracy of determination of concentration: ~5%; spatial resolution of microanalysis $1.0 \times 1.0 \times 3.0$-$5.0 \ \mu m^3$.

The phase composition of the modified layers, that is, the qualitative and quantitative characteristics of the presence of various phases in them, their content, dispersion, structure and chemical composition, in addition to electron diffraction microscopy, was also determined by X-ray phase analysis (XRD-7000s diffractometer, Shimadzu, Japan).

The defective structure of the samples was analyzed by transmission electron diffraction microscopy (TEM) of thin foils (JEM-2100F, JEOL). The JEM-2100F transmission electron microscope has the following main characteristics: a field emission electron source (FEG), which produces a beam of high-brightness electrons (one hundred times brighter than a device with a LaB6 lanthanum hexoboride cathode and much more stable); resolution on points 0.19 nm; resolution on lines 0.4 nm; accelerating voltage varies from 80 to 200 kV; the range of magnifications varies from 50 to 1 500 000 times; the diameter of the electron beam in the transmission electron microscope mode is (2–5) nm. Images of the fine structure of the material were used to classify morphological features of the structure [26].

Foils for studying the structural-phase state of the material by transmission electron diffraction microscopy were prepared by ion thinning of the plates with a thickness of $h \sim 100$ μm, which were cut using the spark method from the sample. The cutting mode was selected in such a way that it did not introduce additional deformation and, therefore, did not affect the structure of the sample.

The plates cut in this way were thinned by the ion etching method (Ion Slicer EM-09100IS device). A distinctive feature of this device is that it does not require the preparation of a disk thinned in the center. Preliminary sample preparation for the Ion Slicer consists only in the manufacture of a parallelepiped with dimensions 2.8×0.5×0.1 mm, which is then closed from a thin wide end with a special protective tape and thinned by a beam of argon ions. The beam energy does not exceed 8 kV, and the angle of incidence can vary from 0° to 6° with respect to the largest face of the sample. This allows one to minimize radiation damage and, thus, preserve the initial structure and phase composition of the sample, and then study them by electron microscopy.

2.4. Methodology of electron–ion–plasma effects

2.4.1. Laboratory installation EVU 60/10 for producing pulsed multiphase plasma jets

Coatings were deposited by pulsed multiphase plasma jets (PMPJ)

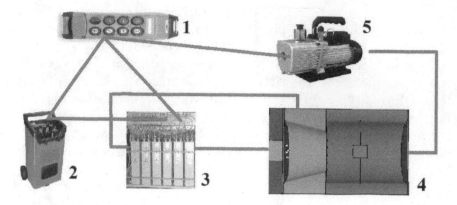

Fig. 2.2. Structural layout of electroexplosive unit EVU 60/10M. 1 – remote control, 2 – charger, 3 – capacity storage, 4 – plasma accelerator and process chamber, 5 – prevacuum pump.

on a laboratory discharge-pulse electric explosive installation EVU 60/10 [27]. Structurally, the electric blasting plant consists of three main parts (Fig. 2.2): a charging device 2, which includes an autotransformer, a step-up transformer and a rectifier; capacitive energy storage 3; a plasma accelerator 4. The installation operates in manual mode – its charge and discharge are performed by pressing the appropriate buttons on the remote control panel 1. The process chamber, in which the multiphase plasma jet is formed, is connected to the foreline pump 5, which is controlled by the remote control 1.

A description of the processes occurring inside the process chamber is given below and shown in Fig. 2.3. When a capacitive energy storage is discharged, high-density electric current flows

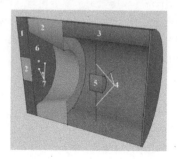
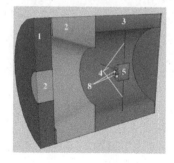

Fig. 2.3. A layout of a pulse plasma accelerator. 1 – insulator, 2 – outward ring and inner cylindrical electrodes, 3 – vacuum process chamber, 4 – holders of samples, 5 – a sample to be treated, 6 – exploded conductor (Al foil), 7 – a weighted Y_2O_3 powder portion, 8 – plasma consisting of atoms Al and Y_2O_3.

through the electrodes 2 through an exploding conductor 6, which leads to its explosion. Explosive products rush into the vacuum process chamber 3 (residual pressure 100 Pa) entraining powder particles, which are used as Y_2O_3. The sample is placed in a vacuum technological chamber at various distances from the nozzle and fixed using sample holders 4. As a result, the products of an electric explosion are a multiphase system including both a plasma component and condensed particles of different dispersion (Y_2O_3), which are deposited on the surface of the workpiece , thus forming a multicomponent coating 8. Leaking onto the surface and reflection from it of the supersonic front of the jet is accompanied by the formation of shock-compressed layer with high values of temperature and pressure. This ensures that the surface is heated to a melting point and higher in a short time of pulsed plasma exposure.

Aluminium foils were used as the material of the exploding conductors, Y_2O_3 was used as a powder sample. The processing was carried out in six modes, differing in discharge voltage, masses of exploding foils and masses of powder samples. All modes used are shown in Table 2.1.

2.4.2. SOLO equipment for processing by intense pulsed electron beam

Surface modification by an intense pulsed electron beam (IPEB) was performed using the SOLO installation, a general view of which is shown in Fig. 2.4. It consists of: a pulsed electron source based on a plasma cathode with a grid stabilization of the plasma boundary; power supply unit of an electronic source; a vacuum technological chamber with a viewing window and a two-coordinate manipulator

Table 2.1. PMPJ modes

Mode No.	Mass of Al foil m_{Al} (mg)	Mass of Y_2O_3 powder $m_{Y_2O_3}$ (mg)	Discharge voltage U (kV)
1	58.9	58.9	2.6
2	58.9	58.9	2.8
3	58.9	29.5	2.6
4	58.9	29.5	2.8
5	58.9	88.3	2.6
6	58.9	88.3	2.8

Fig. 2.4. Appearance of the installation for pulsed surface modification by an electron beam [29].

table, in which a plasma source of electrons and irradiated samples are located; control systems and diagnostics of the parameters of the electron source and electron beam [28].

The installation has the following main advantages over earlier pulsed electronic sources with a plasma cathode: high energy density combined with low accelerating voltage; high energy efficiency; wide range of parameter settings; good reproducibility of pulses; minimum time for preparation; long service life [30].

The electron beam is transported in an axial magnetic field with an argon pressure in the vacuum chamber of 0.01 to 0.1 Pa. The magnetic coils are located at a distance of 17 and 35 cm from each other. Electrons obtained from the emission plasma boundary, stabilized by a grid, fly into a growing magnetic field and then move in a magnetic field located along the coils. When the distance between the coils is less than 17 cm, the resulting electron beam is transferred to the zone of a uniform magnetic field, only slightly changing its diameter (by 3–5%). As the distance between the magnetic coils increases, the electron beam begins to increase its diameter, after which it undergoes re-focusing [29]. Therefore, choosing a different ratio of the current values in the magnetic coils and the distance to the collector, you can adjust the diameter of the electron beam and the energy density at the collector (Fig. 2.5).

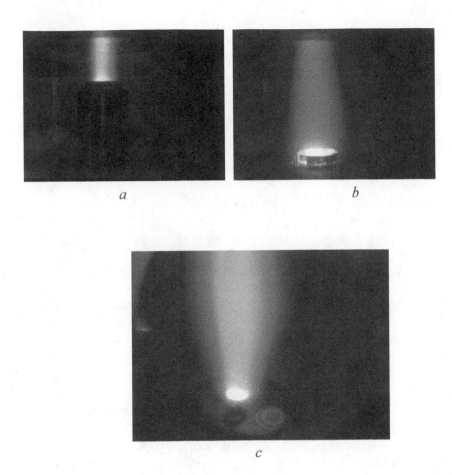

Fig. 2.5. The appearance of the beam in the chamber. The collector is located at a distance of 10 (*a*), 14 (*b*) and 33 cm (*c*) respectively from the drift pipe [30].

When using closely spaced magnetic coils, a more uniform distribution of current on the collector can be obtained, and the location of the collector in the diverging magnetic field (with the distance between the magnetic coils increased to 35 cm) allows increasing the diameter of the beam incident on the collector.

The main parameters of the intense pulsed electron neam (IPEB) of materials that determine the temperature profiles of the heating zone of the surface layers, and, accordingly, the nature and kinetics of structural phase transformations, are the energy density in the electron beam, the duration, number, and repetition rate of irradiation pulses.

In the present work, Al–Si alloy samples were irradiated with an intense pulsed electron beam in 6 modes differing in the electron

Table 2.2. Modes of irradiation of Al–Si alloy samples with a high-intensity pulsed electron beam

Mode No.	1	2	3	4	5	6
Energy density E_s, J/cm²	10	15	20	25	30	35

beam energy density (Table 2.2) and possessing the following identical parameters: accelerated electron energy 17 keV, electron beam pulse duration 150 μs, number of pulses 3, pulse repetition rate pulses 0.3 s^{-1}, residual gas pressure (argon) in the working chamber of the installation $2 \cdot 10^{-2}$ Pa.

2.4.3. Methodology for the complex processing of Al–Si alloy

The first step was the application of the composite coating of the Al–Ti–Y$_2$O$_3$ system by the PMPJ method. In order to increase the intensity of thermal action on the surface of the material prior to its melting and to ensure that the spraying conditions were applied, an end explosion pattern was used. The deposition technology consisted of the following: a two-component Al–Ti foil was clamped between coaxial electrodes, to which an adjustable voltage was applied through a vacuum spark gap. When a capacitive storage device is discharged, an electric current of high density flows through the exploding conductor, which leads to its explosion. Explosive products rush towards the sample being processed, entraining particles of a powder sample, which was used as a powder of Y$_2$O$_3$ (Fig. 2.6 a). As a result, the products of an electric explosion are a multiphase system including both a plasma component (Al, Ti) and condensed particles of different dispersion (Y$_2$O$_3$), which are deposited on the surface of the workpiece, thus forming a multicomponent coating (Fig. 2.6 b). The second stage of processing was the effect of high-intensity pulsed electron beam on the resulting multicomponent coating (Fig. 2.6 c).

The optimal parameters of PMPJ and IPEB, leading to the formation of gradient, multi-element, multiphase, nanostructured states with unique properties in the modified layer, were established for each type of energy impact in [31–33] and are presented in Table 2.3.

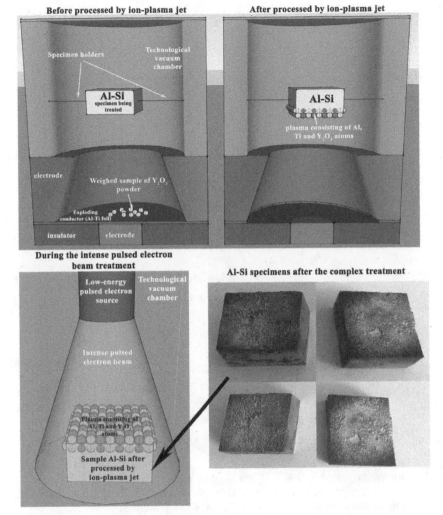

Fig. 2.6. Stages of the complex treatment (*a–c*) and photos of the modified specimens. (d).

Table 2.3. Modes of complex processing of Al–Si alloy

Mode No.	PMPJ parameters			IPEB parameters			
	m (Al-Ti) mg	m (Y_2O_3) mg	U, kV	E_s, J/cm²	E_e, keV	τ, µs	N
1	58.9	58.9	2.8	25	17	150	3
2	58.9	88.3	2.6				
3	58.9	58.9	2.8	25	17	150	3
4	58.9	88.3	2.6				

$m(\text{Al–Ti})$ is the mass of the two-component Al–Ti foil; $m(\text{Y}_2\text{O}_3)$ is the mass of Y_2O_3 powder; U is the discharge voltage; E_s is the electron beam energy density; E_e is the energy of accelerated electrons; τ is the pulse duration of the electron beam; N is the number of pulses.

References for Chapter 2

1. Richards J.W., Aluminium. British Library, Historical Print Editions, 2011.
2. Javidani M., Larouche D., International Materials Reviews. 2014. V. 59. No.3. p. 132–158.
3. Zagulyaev D., Konovalov S., Gromov V., Melnikov A., Shlyarov V., Bulletin of the Polish academy of sciences technical sciences. 2019.V.67. Issue 1. p. 1–5.
4. Konovalov S., Gromov V., Zaguliyaev D., Ivanov Y., Semin A., Rubannikova J., Archives of foundry engineering. 2019.V.19. Issue 2.p. 79–84.
5. Zaguliaev D.V., Konovalov S.V., Ivanov Yu.F., Gromov V.E., Materials Characterization. 2019.V.158. 109934.
6. Zolotorevskiy V.S., Belov N.A., Glazoff M.V., Casting aluminium alloys, first ed. Elsevier Science, 2007.
7. Belov N.A., Belov V.D., Savchenko S.V., Samoshina M.E., Chernov V.A., Alabi A.N., Piston Silumins. Ore and Metals, 2011.
8. Zolotorevsky VS, Belov NA, Metallurgy of cast aluminium alloys. Moscow: MISiS, 2005.
9. Belov NA, Phase composition of aluminium alloys. Moscow: MISiS Publishing House, 2009.
10. Lunev F.A., Silumin. Moscow, Leningrad ONTI, 1937.
11. Laskovnev A.P., Ivanov Yu.F., Petrikova E.A. et al. Modification of the structure and properties of eutectic silumin by electron–ion–plasma treatment. Minsk: Belarus. Science, 2013.
12. Belsky S.E., Volchok I.P., Mityaev A.A., Svidunovich N.A., Casting and metallurgy. 2006. No. 2. S. 130–133.
13. Alsaraeva [Aksenova] K.V., Gromov V.E., Konovalov S.V., Atroshkina A.A., International Journal of Chemical, Molecular, Nuclear, Materials and Metallurgical Engineering. 2015. V.9. No.7. p. 762–766.
14. Ivanov Yu.F., Alsaraeva [Aksenova] K.V., Gromov V.E., Popova N.A., Konovalov S.V., Journal of Surface Investigation. X–ray, Synchrotron and Neutron Techniques. 2015. V.9. No.5. p. 1056–1059.
15. Zeren M., Karakulak E., Journal of Alloys and Compounds. 2008. V. 450. p. 255–259.
16. Hernandez R.C., Sokolowski J.H., Journal of Alloys and Compounds. 2006. V. 419. p. 180–190.
17. Zeren M., Materials and design. 2007. V.28. p. 2511–2517.
18. Taghiabadi R., Ghasemi H.M., Shabestari S.G., Materials Science and Engineering: A. 2008. V. 490. p. 162–167.
19. Li R.X., He L.Z., Li C.X., Materials Letters. 2004. V. 58. p. 2096–2101.
20. Mohamed A. M., Samuel A. M, et al., Materials and Design. 2009. V.30. p. 3943–3957.
21. Zaguliaev D., Konovalov S., Ivanov Y., Gromov V., Applied Surface Science. 2019.V.498. 143767.

22. Schuh C. A., Materials Today. 2007. V.9. No.5. p. 32–40.
23. Mironov VL, Fundamentals of scanning probe microscopy: a textbook for senior students of higher educational institutions. N. Novgorod: RAS, Institute of Physics of Microstructures, 2004.
24. Haviland David B., Current Opinion in Colloid and Interface Science. 2017. V.27. p. 74–81.
25. Inkson B.J., Materials Characterization Using Nondestructive Evaluation (NDE) Methods. Elsevier Ltd, 2016.
26. Prigunova A.G., Belov N.A., Taran Yu.N., Silumins. Atlas of microstructures and fractograms of industrial alloys. Moscow: MISiS Publishing House, 1996.
27. Romanov D.A., Budovskikh E.A., Zhmakin Y.D., Gromov V.E., Steel in translation. 2011. V.41. No.6. p. 464–468
28. Ivanov Yu.F., Ivanov Yu.F., Krysina O. V., Rygina M., Petrikova E.A., Teresov A.D., Shugurov V.V., Ivanova O.V., Ikonnikova I.A., High Temp Material Process. 2014. V. 18. No.4. p. 311–317.
29. Ivanov Yu.F., Koval N.N. The structure and properties of promising metallic materials, ed. Potekaev A.I. Tomsk: NTL Publishing House, 2007.
30. Sosnin K.V., Gromov V.E., Ivanov Yu.F., Structure, phase composition and properties of titanium after electroexplosive alloying with yttrium and electron–beam processing. Novokuznetsk: Publishing house Polygrafist, 2015.
31. Konovalov S.V., Zagulyaev D.V., Ivanov Yu.F., Gromov V.E., Metalurgiya. 2018.V.57. p. 253–256.
32. Zagulyaev D.V., Konovalov S.V., Gromov V.E., Glezer A.M., Ivanov Yu.F., Sundeev R.V., Materials Letters. 2018.V. 229. p.377–380.
33. Konovalov S.V., Zagulyaev D.V., Ivanov Y.F., Gromov V.E., Metalurgiya. 2018.V.57. No.4. p. 253–256.

3

Structural-phase transformations and changes in the properties of Al–Si alloy upon exposure to a pulsed multiphase (Al–Y$_2$O$_3$) plasma jet

3.1. Durometric and tribological studies and metallographic analysis of structural changes in the Al–Si alloy subjected to pulsed multiphase (Al–Y$_2$O$_3$) plasma jet

On the samples subjected to PMPJ (pulsed multiphase plasma jet), microhardness was measured in the sprayed layer and in the substrate at various distances from the treatment surface. Microindentation was carried out on transverse sections. The experimental research results are presented in Table 3.1.

An analysis of the table showed that the microhardness values, both in grains and in the eutectic of modified samples, increase as we approach the sprayed layer. It was found that, regardless of the processing modes, the microhardness of the samples in the alloying zone is greater than at distances of 90 and 70 μm from the edge of the substrate, as well as at the interface between the substrate and the coating (50 μm). The analysis of the dependences gives the basis to conclude that the microhardness of the alloy in the eutectic is greater than in the grains.

Table 3.1. The microhardness test results of samples of Al–Si alloy subjected to PMPJ

Mode No.	Phase	520	90 µm	70 µm	50 µm	Coating
1	Grain	54.00	56.03	51.16	78 48	78.55
	Eutectic	67.04	66.15	72.73	94.17	
2	Grain	63.00	59.36	63.92	87.00	130.51
	Eutectic	73.99	89.5	89.22	87.92	
3	Grain	58.67	58.55	57.53	61.64	57.12
	Eutectic	79.96	76.18	74.78	84.15	
4	Grain	59.53	58.00	67.88	73.34	77.34
	Eutectic	80.37	72.76	80.62	85.65	
5	Grain	63.09	65.61	67.58	66.58	161.2
	Eutectic	108.7	105.17	122.34	111.57	
6	Grain	86.56	66.84	77.00	81.31	101.56
	Eutectic	88.31	94.56	86.49	103.51	

Studies of changes in the microhardness value with distance from the processing surface showed that the optimal characteristics of the resulting treated surface, which were the thickness and microhardness of the obtained layer, are observed for the processing modes 2 and 5 (Fig. 3.1 *a*).

Analysis of Fig. 3.1 *a* shows that the microhardness of the material is maximum in the surface layer and exceeds the microhardness of the cast alloy by more than 2 times. As the distance from the surface of the modification increases, the microhardness decreases and reaches a microhardness of the initial Al–Si alloy at a depth of ≈90 µm.

The thickness of the sprayed layer of the sample treated in mode 2 was 50 µm (Fig. 3.1 *b*), and in mode 5 it was 80 µm.

When analyzing the cross-sectional structure of the samples treated according to the modes 2 and 5, the formation of a multilayer structure was revealed, which consists of a highly porous coating, inhomogeneous in thickness (1), a layer of liquid-phase alloying (2), and a layer of thermal influence (3) (Fig. 3.2 *a* , *b*). The thickness of the modified layer varies between 17–117 µm for a sample treated in mode 2 and 33–60 µm for a sample processed in mode 5, which is due to the selection of the mass of the sprayed powder, where for mode 2 the mass of Y_2O_3 is less than for mode 5. Thus, the coating of the samples treated in mode 5 turned out to be more uniform in width than in the samples processed in mode 2. The thickness of

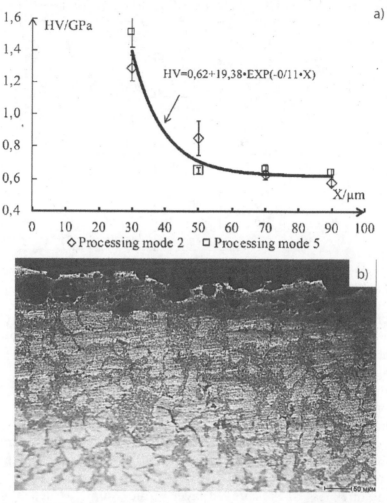

Fig. 3.1. a) Microhardness profile of silumin after electroexplosive alloying (the load on indenter 0.50 N). b) The coating of the Al–Y$_2$O$_3$ system produced by the method of electroexplosive alloying (×20, optical microscopy, processing mode 2).

the heat-affected layer was 150–260 µm for mode 2 and 53–80 µm for mode 5.

In the sample processed in the second mode, the thermal influence layer is much wider than during the treatment with mode 5, which is due to the selection of a larger value of the discharge energy. Nanohardness measurements using the SUPER NANOSCAN scanning nanometerhardness meter by sclerometry were carried out on samples processed according to one of the optimal modes – mode No. 2.

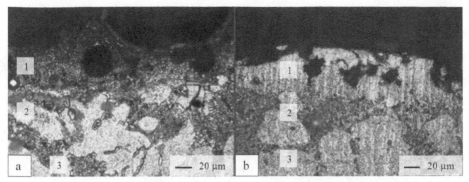

Fig 3.2. Optical microscopy of the surface profile of Al–Si alloy samples with processing modes 2 (*a*) and 5 (*b*).

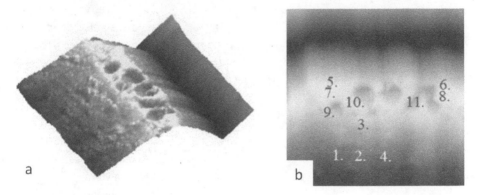

Fig. 3.3. a) Three-dimensional surface profile map (Zopt) for force probe scanning of the sample surface; b) diagram of the distribution of points of measurement of nanohardness on the test surface of a sample.

Studies of the structure obtained using the optical attachment to the NanoScan device revealed that the surface layers of the sample processed according to mode 2, i.e. spraying areas have a double structure. The layer located closer to the surface is porous, and when moving deeper into the product, a continuous layer is observed. The pore structure in the surface layer of the coating is shown on a three-dimensional map of the surface profile (Zopt) during force probe scanning of the surface (Fig. 3.3 *a*).

To analyze the changes in mechanical properties along the surface of the studied area, the technique of restoring the supply curves at given points was used. Nanohardness was selected as a value that determines the mechanical properties. The oscillation frequency of the probe was 450 Hz. From points 3 to 7 measurements were made to each selected area, which allows one to reliably determine the values of nanohardness. The distribution of the measuring points is shown in Fig. 3.3 *b*,

The measured nanohardness values are presented in Table 3.2. An analysis of the nanohardness values at various points of the studied area showed that a series of points 1 and 4, presumably, refers to grains of an α-solid solution of the basic material and, on average, corresponds to a value of 1.9 GPa. Point 2 according to the value of nanohardness corresponds to the eutectic site and is also supposedly located outside the coverage area. Point 3 refers to a continuous coating layer, where the nanohardness value above is averaged between the values of the grains of the α-solid solution and the eutectic.

As we move toward the porous coating layer, the values of nanohardness decrease by 24% relative to the dense layer. For example, at point 9, it is believed that there is a transition zone from a continuous to a porous coating layer. Based on the measurement values at points 10 and 11, the nanohardness of the porous coating

Table 3.2. Nanohardness values depending on the studied site

Measurement No.	Nanohardness, GPa	Comment
1	1.97	Grains of α-solid solution, outside coating zone
2	2.39	Eutectic section, mostly outside the coating zone
3	2.06	Continuous coating layer
4	1.81	Grains of α-solid solution, outside coating zone
5	1.20	Porous coating layer, section away from pore
6	1.74	Porous coating layer, section away from pore
7	1.50	Porous coating layer, section away from pore
8	1.90	Porous coating layer, section away from pore
9	1.56	Transition zone from continuous coating layer to porous layer
10	1.29	Porous coating layer, section close to pore
11	1.37	Porous coating layer, section close to pore

layer is, on average, 1.3 GPa, this is due to the effect of pores on the mechanical properties of the material. However, the values of nanohardness at points 6, 7, 8, located at some distance from the defects reaches an average of 1.7 GPa.

As a result of the studies using a scanning probe nanometer hardness tester, it was revealed that the grain structure of the main material is α-solid solution and eutectic grains, the coating depth varies from 40 to 60 μm, which correlates with data from metallographic analysis.

The tribological properties of the modified Al–Si alloy layer were studied on samples treated according to mode 2, determining the wear resistance and friction coefficient. The change in the friction coefficient during the testing of the material for wear resistance is shown in Fig. 3.4. An analysis of the results shows that the wear resistance of the studied material after PMPJ increased, compared with the wear resistance of the initial Al–Si alloy, by more than 28 times; the friction coefficient was more than halved.

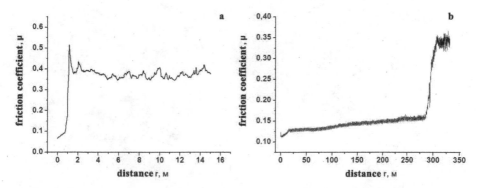

Fig. 3.4. Dependence of the value of the friction coefficient on the distance travelled by the counterbody along the friction track during wear tests; a – Al–Si alloy in a cast state; b – Al–Si alloy after exposure to PMPJ.

Thus, the impact of the PMPJ containing particles of the yttrium oxide powder on the surface of the eutectic Al–Si alloy is accompanied by the formation of a surface layer whose mechanical (microhardness) and tribological (wear resistance and friction coefficient) properties are many times higher than the corresponding characteristics of the Al–Si cast alloy.

As a result of studying changes in durometric properties and performing a metallographic analysis of structural changes, it was found that the optimal of all used processing modes are the modes

2 and 5 (Table 2.1). Further studies using atomic force microscopy, scanning and transmission electron microscopy, and X-ray diffraction analysis were performed on a sample processed according to optimal conditions.

3.2. Study of the morphology of the Al–Y$_2$O$_3$ using atomic force microscopy

The study of the coating and the heat-affected zone of samples of the Al–Si alloy treated with PMPJ, according to mode 2, was carried out using an NT-MDT Solver NEXT atomic force microscope, since this method is one of the most powerful modern methods for studying the morphology and local surface properties of solids with high spatial resolution. Pictures of the surface profile topography were taken. Surface profile topography study mode: semi-contact mode, probe type NC-HA.

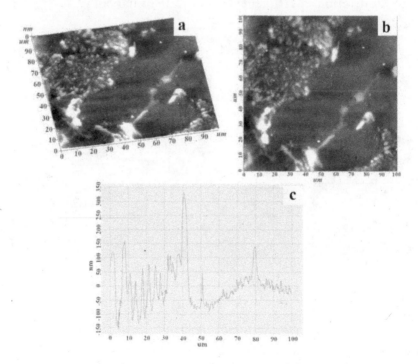

Fig. 3.5. Atomic force microscopy of the surface profile of the bulk of the sample (mode 2): *a* – distribution of the surface roughness in height in 3D format, *b* – 2D image of the topography of the surface profile, with secant section, *c* – distribution of the roughness along the base length.

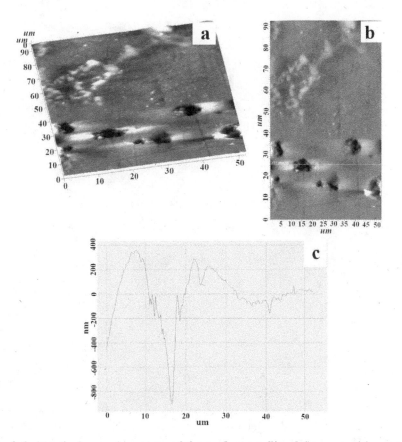

Fig. 3.6. Atomic force microscopy of the surface profile of the sprayed layer of the sample (mode 2): *a* – distribution of the surface roughness in height in 3D format, *b* – 2D image of the surface profile topography, with secant section, *c* – distribution of the roughness along the base length.

The topography of the surface profile of the samples obtained using atomic force microscopy is shown in Figs. 3.5–3.8. 3D and 2D images are shown, as well as sections of the roughness distribution along the baseline in the bulk of the samples and on the coating.

The structure of the sample of the modified Al–Si alloy in the main volume is represented by grains reaching a height of 350 nm and, relatively, a smooth eutectic.

From Fig. 3.6 we can conclude that the coating of the sample treated according to mode 2, is highly porous. Statistical processing of the image revealed a maximum pore depth, which has a value of the order of 1500 nm. The pore size reaches up to 10 μm in diameter.

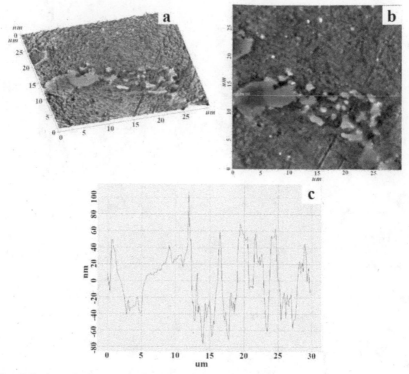

Fig. 3.7. Atomic force microscopy of the surface profile of the bulk of the sample (mode 5): *a* – distribution of the surface roughness in height in 3D format, *b* – 2D image of the surface profile topography, with a secant section, *c* – distribution of the roughness along the base length.

The main volume of the sample processed according to mode 5, as well as the main volume of the sample processed according to mode 2, has a structure consisting of eutectic and grains.

From Fig. 3.8 it can be seen that the coating in the sample treated according to mode 5 is porous. The number and depth of pores is much smaller than in the sprayed layer of sample 2. The maximum pore depth reaches 500 nm. This can be explained by the fact that, due to the greater weight of the weight of the sprayed powder and less energy impact, the particles of the plasma jet, having lower energy, were distributed more evenly. A possible fact is the occurrence of Rayleigh–Taylor instability, that is, mixing of the sprayed and fused layers, as a result of which the number of pores turned out to be less than for a sample treated according to mode 2.

Investigations of the structure of the surface profile of the Al–Si alloy by atomic force microscopy showed that PMPJ leads to the

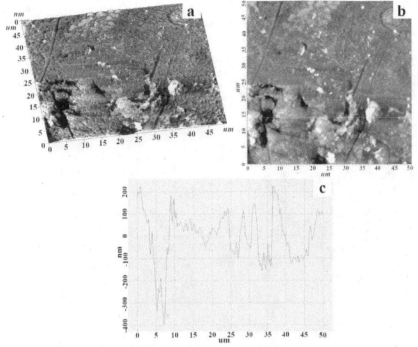

Fig. 3.8. Atomic force microscopy of the surface profile of the sprayed layer of the sample (mode 5) *a* is the distribution of the surface roughness in height in 3D format, *b* is a 2D image of the topography of the surface profile, with a secant section, *c* is the distribution of the roughness along the base length.

formation of a multilayer structure, which consists of a highly porous coating, inhomogeneous in thickness, a layer of liquid-phase alloying and a heat-affected layer.

3.3. Study of the phase structure and surface morphology of the Al–Si alloy modified by the Al–Y$_2$O$_3$ system

3.3.1. Analysis of the structure of the alloy in the initial state

The Al–Si alloy, which is the material of the study of this work, contains copper, magnesium, iron, nickel and some other alloying elements. The presence of alloying elements contributes, on the one hand, to an increase in strength, and on the other hand, leads to a decrease in crack resistance, which is due to the formation of intermetallic lamellar morphology [29–32]. Typical images of the structure of the etched thin section obtained by scanning electron

54

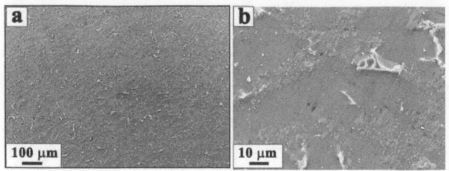

Fig. 3.9. Alloy structure revealed by scanning electron microscopy of etched thin section.

microscopy and demonstrating the multiphase, morphologically diverse nature of the material are presented in Fig. 3.9.

Using X-ray microspectral analysis methods (mapping method), it was established that the main alloying elements are nickel, magnesium and copper (Fig. 3.10).

X-ray microanalysis allows the study of the elemental composition of a locally specifically selected inclusion. The results of the analysis

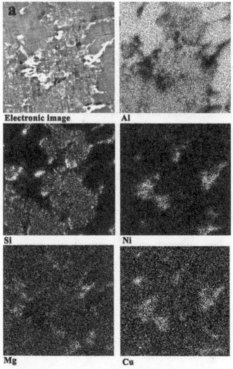

Fig. 3.10. Maps of the distribution of alloying elements over the area of the initial alloy sample, the electron-microscopic image of which is shown in (*a*).

X-ray microanalysis allows the study of the elemental composition of a locally specifically selected inclusion. The results of the analysis of local regions are presented in Fig. 3.11 and in Table 3.3. It is clearly seen that the alloying elements of the alloy are distributed very nonuniformly in the material, forming compounds that differ in size, contrast level, morphology, and elemental composition.

The phase composition of the investigated Al–Si alloy was determined by X-ray diffraction analysis. The plot of the x-ray obtained from the test material is shown in Fig 3.12. The results of the quantitative analysis of the X-ray are given in Table 3.4.

Analyzing the results presented in the table, it can be noted that the main phases of the studied material, as one would expect, are solid solutions based on aluminium and silicon. The crystal

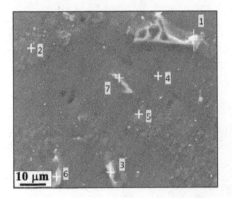

Fig. 3.11. Electron microscopic image (SEM) of the structure of the investigated alloy; areas in which X-ray microanalysis of the elemental composition of the material was performed are indicated.

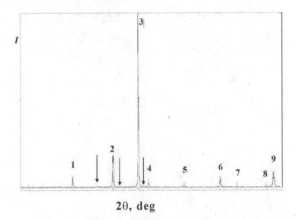

2θ, deg

Fig. 3.12. X–ray plot of the investigated Al–Si alloy; the numbers indicate the diffraction maxima of aluminium and silicon: 1 – (111) Si; 2 – (111) Al; 3 – (200) Al; 4 – (220) Si; 5 – (311) Si; 6 – (220) Al; 7 – (400) Si; 8 – (331) Si; 9 – (311) Al. The arrows indicate the diffraction lines of the $AlCu_3$ phase.

Table 3.3. The results of X-ray microspectral analysis of the surface area of the alloy, the electron-microscopic image of which is shown in Fig. 3.11. The results are presented in weight. %

Region	Si	Ni	Cu	Fe
	0.60	13.5	13.30	0.00
	8.70	0.30	2.20	0.00
	1.70	11.8	14.00	0.00
	0.50	0.20	1.30	0.00
	22.5	1.10	1.60	1.20
	1.1	14.8	15.8	0.5
	2.3	17.2	5.2	2.7

Table 3.4. Results of X-ray diffraction analysis of a sample of an Al–Si alloy in a cast state. The results are presented in wt. %

Phase	Content, rel.%	Type of lattice	Lattice parameter, nm		Atomic radius, nm
			a_0	a	
Al	84.2	Fm3m	0.4050	0.40484	0.143
Si	12.3	Fm3ms	0.54307	0.54265	0.132

Comment: a_0 – tabulated value; a – value in the alloy

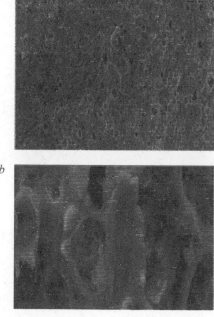

a

b

Fig. 3.13. Electron microscopic image (SEM) of the structure of the investigated alloy after PMPJ.

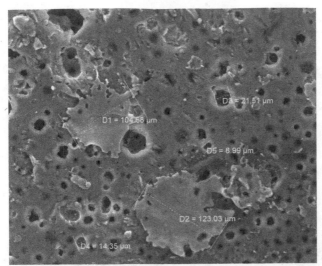

Fig. 3.14. The dimensions of the residues of the coating not evaporated during the processing with PMPJ.

lattice parameters of aluminium and silicon in the alloy under study are close to the crystal lattice parameters of pure elements, which indicates the stratification of these elements during crystallization of the alloy.

3.3.2. Analysis of the structure of the alloy of modified PMPJ (m_{Al} = 58.9 mg, $m_{Y_2O_3}$ = 58.9 mg, U = 2.8 kV)

The studies performed in this work showed that as a result of the PMPJ of the alloy under study, a surface layer is formed, characterized by a high level of roughness, containing a large number of micropores, microcraters and microcracks (Fig. 3.13).

In some places on the surface, the remnants of the coating that did not evaporate during the PMPJ process are observed (Fig. 3.14).

Table 3.5. The element composition of the surface layer in Al–Si alloy after pulsed multiphase plasma jet, identified by the methods of X-ray microspectral analysis of zones shown in Fig. 3.16, wt. %

Zone	Al	Si	Mg	Ti	Fe	Ni	Cu	Y	O	C
Fig. 3.16a	47.2	3.0	0.6	1.0	0.7	1.3	1.8	16.2	10.8	17.4
Fig. 3.16b	0.8	0.0	0.0	0.2	0.7	0.5	0.7	34.0	28.1	35.0

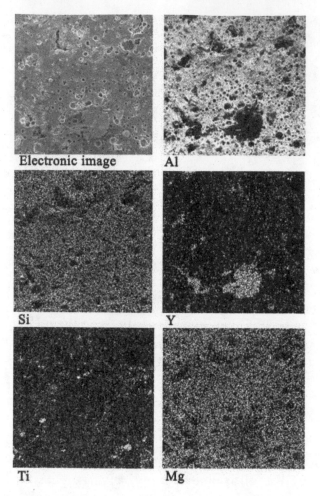

Fig. 3.15. The distribution of elements on the surface of the sample after PMPJ.

The dimensions of the craters formed as a result of processing vary from 5 to 20 μm. The dimensions of the preserved coatings reach 100–130 μm.

Mapping the surface of the sample indicates the heterogeneity of the distribution of elements (Fig. 3.15).

The residues of the coating that did not evaporate during the exposure process contain up to 65% yttrium with an oxygen content of up to 20–26%. The porous surface of the sample is oxidized (the amount of oxygen reaches 5–6%). The depth of porosity formation near the surface after treatment with PMPJ is 40–50 μm (Fig. 3.15).

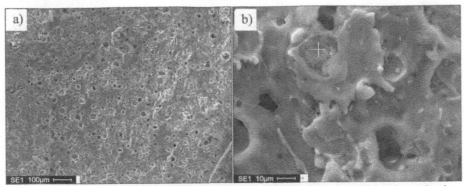

Fig. 3.16. SEM data on the structure of Al–Si alloy surface treated by the pulsed multiphase plasma jet: a), b) see Table 3.5.

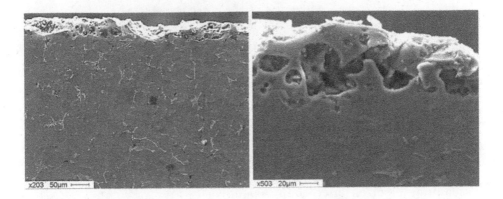

Fig. 3.17. SEM data on the structure of Al–Si alloy surface treated by the pulsed multiphase plasma jet (cross section),

X-ray spectral analysis methods were used to study the elemental composition of the surface layer of the alloy subjected to PMPJ. The results of the studies are presented in Fig. 3.16 and in Table 3.5.

Analyzing the results given in Fig. 3.16 and Table 3.5, it should be noted that after pulsed multiphase plasma jet the surface layer is formed with a high level of alloying elements distributed non-homogeneously, it is most visible in spreading of yttrium and oxygen atoms. To be more precise, there are zones with concentration of yttrium and oxygen exceeding the average one twice and more. These results confirm that in the plasma flow of powder particles there is an alloying material identified in some studies before. Furthermore, the surface layer with a high level of roughness

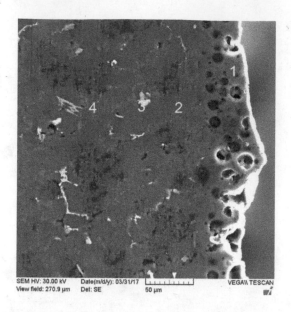

Fig. 3.18. Alloy areas selected for analysis of elemental composition.

containing a lot of micropores, micro-craters and microcracks is formed in electroexplosive alloying.

The structure of the volume of the alloy subjected to treatment with PMPJ was analyzed by the method of transverse sections. A typical image of the structure of the modified layer revealed by scanning electron microscopy is shown in Fig. 3.17. Analyzing the presented results, it can be noted, firstly, that the thickness of the modified layer varies in the range (30–50) μm. Secondly, a high level of porosity. Pores are located throughout the thickness of the modified layer; pore sizes vary from units to tens of micrometers.

X-ray spectral analysis methods were used to study the elemental composition of the volume of the studied material on transverse sections exposed to PMPJ at different points of the sample (Fig. 3.18).

Table 3.6. Elemental composition in different areas of the sample / wt. %

Region	Mg	Al	Si	Ti	Fe	Ni	Cu
1	00.96	84.62	10.51	00.20	00.28	01.18	02.26
2	00.17	86.09	12.19	00.06	00.00	00.06	01.43
3	00.40	54.23	00.77	00.00	00.48	19.63	24.47
4	00.64	67.41	29.94	00.00	00.20	00.19	01.62
5	00.11	74.73	00.64	00.00	00.69	15.74	08.08

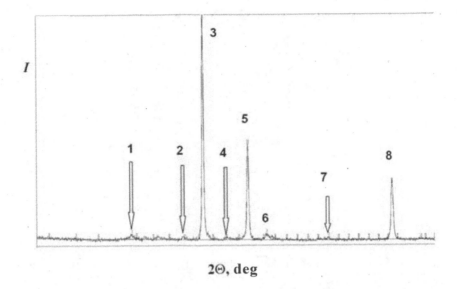

2Θ, **deg**

Fig. 3.19. X-ray plot of the investigated Al–Si alloy subjected to modification with PMPJ; the numbers indicate the diffraction maxima of aluminium, silicon, and Y_2O_3: 1 – (111) Si; 2 – (411) Y_2O_3; 3 – (111)Al; 4 – (422)Y_2O_3; 5 – (200)Al; 6 – (220) Si; 7 – (311)Si; 8 – (220)Al.

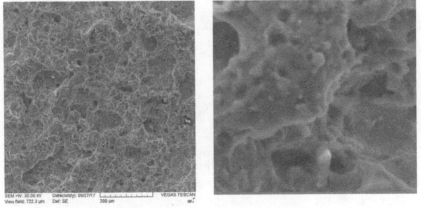

Fig. 3.20. The surface of the samples after the effect of PMPJ.

The results of the analysis of local areas of the material marked in Fig. 3.18 are presented in Table 3.6.

The phase composition of the Al–Si alloy modified by PMPJ was also investigated by X-ray diffraction analysis. Figure 3.19 shows the x-ray obtained from the studied material. Quantitative results of the phase analysis of the material are presented in Table 3.7. Analyzing the results presented in Table 3.7, one can note, firstly, a substantially high level of silicon in the surface layer of the alloy,

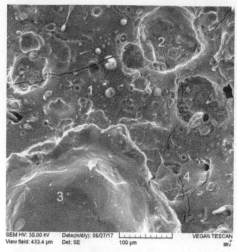

Fig. 3.21. Zones with a sharp heterogeneity of the Al–Si alloy structure after exposure to PMPJ.

which may indicate the evaporation of a certain layer of aluminium during PMPJ. Secondly, the presence of the Y_2O_3 phase, which may be due to the introduction of particles of the initial yttrium oxide powder into the surface layer of the material being modified.

3.3.3. Analysis of the structure of the alloy of modified with PMPJ (m_{Al} = 58.9 mg, $m_{Y_2O_3}$ = 29.5 mg, U = 2.6 kV)

A general view, with a different increase, of the surface of an Al–Si alloy sample after processing in the second optimal mode (Table 2.1, mode 5) is shown in Fig. 3.20. The average values of the content of various elements in the surface layer are shown in Table 3.8.

Analyzing the presented results, it can be stated that the surface of the sample is heterogeneous in chemical composition, covered with craters and pores. Inside the craters, the remains of the coating film are visible on the surface. In some places of the surface subjected to modification, zones with a sharp heterogeneity of the structure and

Table 3.7. Elemental composition in different zones of the sample (CSR – coherent scattering region)

Phase	Content, rel.%	Lattice parameter, nm	D(CSR), nm	$\Delta d/d$, 10^{-3}
Al	68.2	0.40485	75.01	0.24
Si	30.1	0.54231	16.40	0.80
Y_2O_3	01.7	1.06010	16.60	7.88

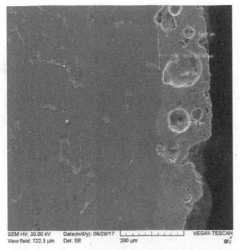

Fig. 3.22. Microstructure of a transverse section of an Al–Si alloy after exposure to PMPJ.

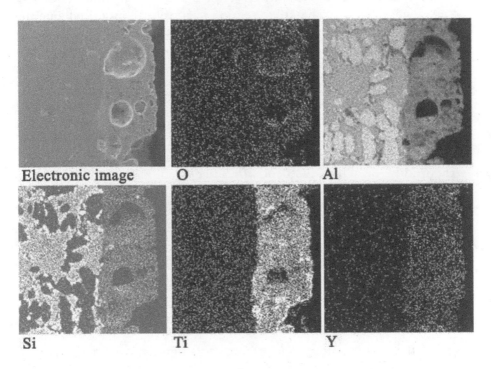

Fig. 3.23. Microstructure of a transverse section of an Al–Si alloy after exposure to PMPJ.

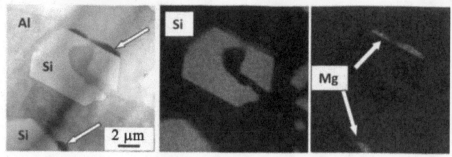

Fig. 3.24. X-ray microanalysis of the elemental composition of the Al–Si alloy (CTEM method).

a large number of cracks formed as a result of high-speed cooling are observed (Fig. 3.21).

The elemental composition of the individual zones of the sample (Fig. 3.21) is presented in Table 3.9; we can state the heterogeneity of the distribution of elements over the surface of the modification.

A general view of the cross section surface of a modified sample is shown in Fig. 3.22. The coating thickness is about 155 μm, there are a large number of micropores of various diameters from 32 μm to 72 μm. The chemical composition of the modified layer is enriched with Al, Y, and Ti atoms (Fig. 3.23), these elements coincide with the components of a pulsed multiphase plasma jet, which indicates that the layer under consideration belongs to the sprayed coating. At distances greater than 155 μm, the structure becomes relatively uniform. Most likely, this is a substrate of the Al-Si alloy, on which the deposition occurred, and this is indicated by individual intermetallic inclusions, which are easily identified by the distribution maps of chemical elements shown in Fig. 3.23.

3.4. Phase transformations of the surface layer of an Al–Si alloy subjected to a pulsed multiphase plasma jet

3.4.1. Studies of the morphology and elemental composition of the Al–Si alloy in the initial state

The initial morphology and elemental composition of the Al–Si

Table 3.8. The average chemical composition of the sample zone shown in Fig. 3.20. The results are presented in wt. %

O	F	Mg	Al	Si	S	Ti	V	Fe	Ni	Cu	Y
17.25	1.77	0.39	33.04	3.69	0.63	7.45	0.35	0.50	3.10	1.23	30.60

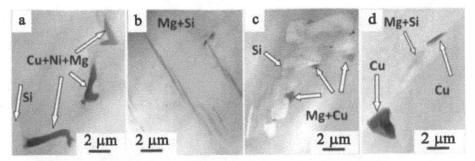

Fig. 3.25. Electron microscopic image of the structure of an Al–Si alloy obtained in the CTEM analysis mode when studying the elemental composition of the material.

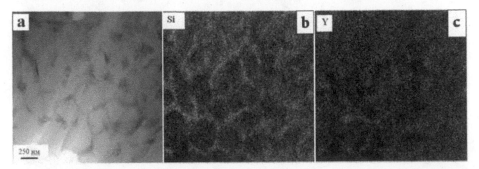

Fig. 3.26. X-ray microspectral analysis (mapping method) of the elemental composition of an Al–Si alloy subjected to PMPJ (CTEM method); *a* is a bright field image; *b* is the image obtained in x-ray radiation of silicon atoms; *c* is the image obtained in the x-ray radiation of yttrium atoms.

the Al–Si alloy in the initial state

The initial morphology and elemental composition of the Al–Si alloy was analyzed by transmission electron microscopy using an attachment

Figure 3.24 shows the results of a study of the distribution of silicon and magnesium atoms in the studied alloy, revealed by STEM

Table 3.9. The chemical composition of the different zones of the sample marked in Fig. 3.21. The results are presented in weight %

Zone	O	F	Mg	Al	Si	Ti	V	Fe	Ni	Cu	Y	Mo
1	18.49	1.84	0.48	43.26	4.04	13.26	0.56	0.42	5.25	1.63	9.87	0.88
2	18.76	2.80	0.17	15.06	1.53	7.07	0.21	0.37	4.60	1.21	47.22	1.00
3	19.35	1.90	0.18	10.85	1.08	6.89	0.36	0.51	7.03	4.20	45.31	2.34
4	27.96	1.47	0.11	12.90	2.30	8.83	0.22	0.24	2.32	0.51	42.46	0.68

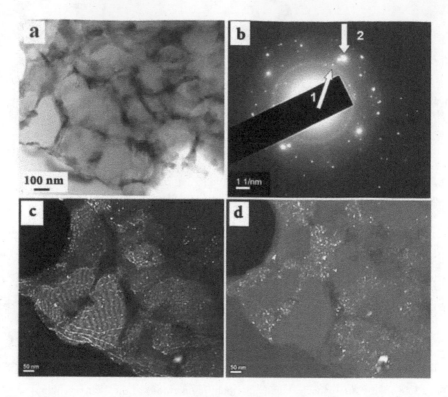

Fig. 3.27. Electron microscopic image of the structure of an Al–Si alloy subjected to PMPJ; *a* is the bright field; *b* – microelectron diffraction pattern; arrows indicate reflexes in which dark fields are obtained; *c* is the dark field obtained in the reflexes indicated by 2[111] Al + [112] YSi$_2$; *d* is the dark field obtained in reflections 1 [111] Si + [111] Y$_2$Si$_2$O$_7$ (ring).

methods. The elemental composition of the foil region shown in Fig. 3.24, is as follows (at.%): 74.2 Al, 23.6 Si, 1.4 O, 0.6 Mg, 0.2 Fe.

Figure 3.25 shows other distribution options for the atoms of alloying elements contained in the alloy under study, identified by X-ray spectral analysis (STEM method).

Thus, it has been shown by SEM and STEM methods that the alloying elements present in the alloy under study are distributed nonuniformly, forming inclusions having a diverse composition, shape and size.

The Al–Si alloy studied in this work is a multiphase material and contains, in addition to phases based on solid solutions of aluminium and silicon, intermetallic compounds of various compositions. Particles of silicon and intermetallic compounds have a diverse shape (globular, lamellar, needle-shaped or skeletal) and, therefore, can

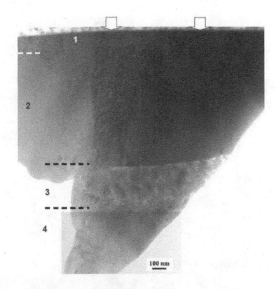

Fig. 3.28. Electron microscopic image of the structure of the Al–Si–Y–O alloy. The arrows indicate the surface of the impact of PMPJ, the numbers indicate the numbers of the identified layers.

not only strengthen the material, but also be sources of microcracks during the operation of products.

3.4.2. Studies of the morphology and elemental composition of the modified surface of the Al–Si alloy

The defective substructure, elemental and phase composition of the surface layer of the Al–Si alloy subjected to PMPJ were also analyzed by transmission electron diffraction microscopy of thin foils. It has been established that as a result of high-speed cooling of the modified layer that occurs during PMPJ, a cellular crystallization structure of aluminium is formed in the surface layer (Fig. 3.26). The size of the crystallization cells varies in the range of (200–450) nm. At the boundaries of the cells are interlayers of the second phase. Using X-ray microspectral analysis, it was found that the interlayers are formed by silicon and yttrium atoms.

The results of a quantitative analysis of the elemental composition of the sample site shown in Fig. 3.26a are given in Table 3.10. A low concentration of oxygen atoms is noteworthy, which may indirectly indicate the absence of an oxide phase at the boundaries of crystallization cells.

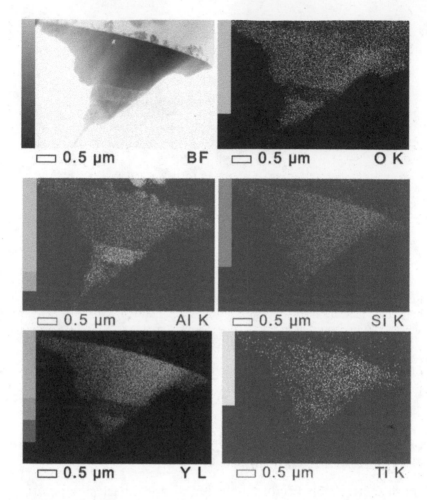

Fig. 3.29. Distribution maps of atoms of elements forming the surface layer of a portion of a sample of an Al–Si alloy, the electron-microscopic image of which is shown in (a) exposed to PMPJ.

The phase analysis of the surface layer of the Al–Si alloy subjected to PMPJ was studied by transmission diffraction electron microscopy using the dark-field analysis technique and displaying the corresponding microelectron diffraction patterns (Fig. 3.27).

Analysis of Fig. 3.27 shows that the surface layer of the alloy subjected to PMPJ has a cellular crystallization structure. It is clearly seen that the cells are oval. Cell sizes vary from 150 nm to 350 nm. Microdiffraction analysis showed that the cells are an aluminium-based solid solution (Fig. 3.27 *d*). Particles of the second phase are detected in the volume of cells by dark-field analysis

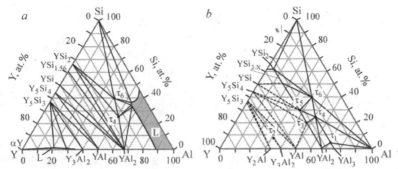

Fig. 3.30. Isothermal sections of the ternary Al–SiY system at different temperatures: $T = 1025°C$ (*a*) and $T = 500°C$ (*b*) [2, 3].

Table 3.10. The results of X-ray microspectral analysis of the elemental composition of the Al–Si alloy subjected to PMPJ (wt.%, Al – the rest)

Si	Mg	Ni	Fe	Y	O
6.8	1.4	1.0	0.5	1.5	0.5

methods; particle sizes vary within units of nanometers (Fig. 3.27 *c*). An analysis of the microelectron diffraction pattern (Fig. 3.27 *b*) suggests that these particles are yttrium silicides of the composition YSi_2. The aluminium cells are separated by interlayers of the second phase. An analysis of the microelectron diffraction pattern (Fig. 3.27) shows that the interlayers are formed by silicon and yttrium silicate with the composition $Y_2Si_2O_7$.

Analysis of Fig. 3.27 shows that the surface layer of the alloy subjected to PMPJ has a cellular crystallization structure. It is clearly seen that the cells are oval. Cell sizes vary from 150 nm to 350 nm. Microdiffraction analysis showed that the cells are an aluminum-based solid solution (Fig. 3.27 *d*). Particles of the second phase are detected in the volume of cells by dark-field analysis methods; particle sizes vary within units of nanometers (Fig. 3.27 *c*). An analysis of the microelectron diffraction pattern (Fig. 3.27 *b*) suggests that these particles are yttrium silicides of the composition YSi_2. The aluminum cells are separated by interlayers of the second phase. Analysis of the microelectron diffraction pattern (Fig. 3.27) shows that the interlayers are formed by silicon and yttrium silicate with the composition $Y_2Si_2O_7$.

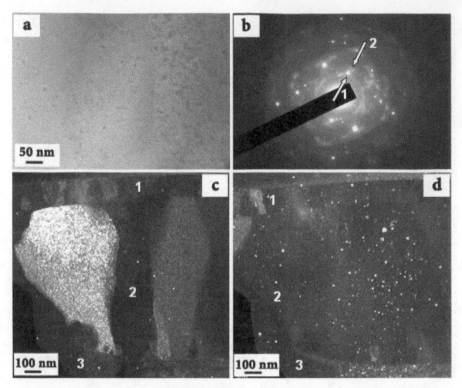

Fig. 3.31. Electron microscopic image of the structure of layers 1 and 2 indicated in Fig. 3.28; *a* is a bright field image of layer 2; *b* – microelectron diffraction pattern of plot (*a*); *c, d* are the dark fields obtained in the [111] Y_2O_3 reflexes (reflex 1 to (*b*)) and [031] Al_2SiO_5 (reflex 2, to (*b*)), respectively. The numbers on (*c*) and (*d*) indicate the layers, respectively Fig. 3.28.

3.4.3. Studies of the multilayer structure and elemental composition of the modified surface layer of Al–Si alloy

A typical electron microscopic image of the surface layer of an Al–Si alloy subjected to PMPJ is shown in Fig. 3.28. It is clearly seen that a multilayer structure is formed, represented by a surface layer (layer 1), intermediate layers (layer 2 and layer 3) and a transition layer (layer 4). The longest of the identified layers is layer 2, the thickness of which is ≈900 nm, the thinnest is layer 1 – ≈100 nm. The formation of a multilayer structure may be due to several factors: a gradient of the temperature field, variation in the elemental composition, a gradient in the cooling rate, etc.

The elemental composition of the surface layer of the Al–Si alloy shown in Fig. 3.28 was determined by X-ray spectral analysis methods. The analysis results are shown in Table 3.11. As a result

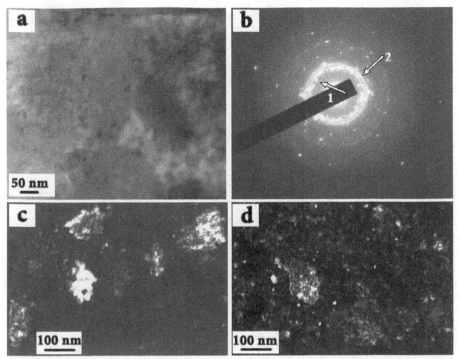

Fig. 3.32. Electron microscopic image of the structure of layer 3 indicated in Fig. 3.28; *a* is a bright field image; *b* – microelectron diffraction pattern of image (*a*); *c, d* are the dark fields obtained in the reflections of [200] Al_2O_3 (reflex 1 to (*b*)) and [131] Al_2SiO_5 (reflex 2, to (*b*)), respectively.

of the studies, it was found that the analyzed volume of the foil is a multi-element alloy, the main elements of which are oxygen, aluminium, yttrium silicon and titanium. The mapping method revealed the preferred location of these elements in the analyzed material. Presented in Fig. 3.29 mapping results show that the oxygen and yttrium atoms are preferably located in layers 1, 2 and 4; aluminium atoms in layer 3; silicon and titanium atoms in layers 1–3.

The phase composition and defective substructure shown in Fig. 3.28. layers were analyzed by diffraction electron microscopy. Anticipating the research data, we carried out an analysis of the phase formation taking place in the Al–Si–Y system under equilibrium conditions. The Al–Si system is of a simple eutectic type with a small solubility of the components in each other in the solid state [1]. The maximum solubility of Si in solid (A1) is observed at a eutectic temperature of 577°C and is equal to ~1.5 ± 0.1 at.%. Therefore, in the Al–Si system, we can talk

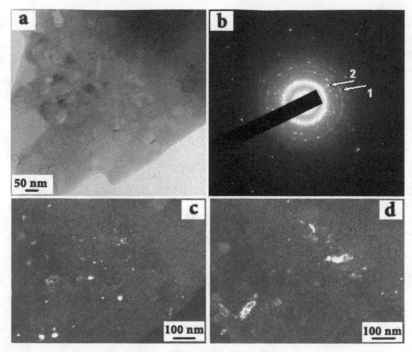

Fig. 3.33. Electron microscopic image of the structure of layer 4 indicated in Fig. 3.28; a – bright field image; b – microelectron diffraction pattern of image (a); c, d are the dark fields obtained in the reflections of [224] Y_3Al_2 (reflex 1 to (b)) and [311] $Y_2Si_2O_7$ (reflex 2 to (b)), respectively.

about the existence of an aluminium-based solid solution. The maximum solubility of Al in (Si) is very small and equal to 0.016 ± 0.003 at.% at a temperature of 1190°C, which indicates the presence of practically pure silicon crystals in the system.

The Al–Y system contains five intermetallic compounds (Al_3Y, Al_2Y, AlY, Al_2Y_3, and AlY_2), which are formed as a result of various

Table 3.11. X-ray microanalysis of the foil portion shown in Fig. 3.28

Element	Excitation energy, keV	Relative content, at.%
O (K)	0.525	49.80
Al (K)	1.486	27.18
Si (K)	1.739	00.60
Ti (K)	4.508	01.27
Fe (K)	6.398	00.25
Ni (K)	7.471	00.19

reactions [1]. Compounds Al_2Y and Al_2Y_3 melt congruently at temperatures of 1485°C and 1100°C, respectively. Compounds Al_3Y, AlY, and AlY_2 are formed by peritectic reactions at temperatures of 980°C, 1130°C, and 985°C, respectively. Two eutectic reactions take place in the system: $L \leftrightarrow (Al) + Al_3Y$ and $L \leftrightarrow (Y) + AlY_2$ at temperatures of 639°C and 960°C, respectively. The solubility of Y in Al is low and amounts to 0.052 at.% [1].

Five intermetallic compounds have been established in the Si–Y system: Y_5Si_3, Y_5Si_4, YSi, $YSi_2–X$ ($YSi_{1.56}$), and YSi_2 [1]. Of these compounds, only one, Y_5Si_3, melts congruently at 1875°C. The compound YSi_2 is metastable. In the Si–Y system two eutectic transformations are observed: $L \leftrightarrow Y_3Si_5 + (Si)$ and $L \leftrightarrow (\pm Y) + Y_5Si_3$ at temperatures of 1230°C and 1260°C, respectively [1].

In the Al–Si–Y system of three alloys of the forming elements, silicon is referred to as non-metals, since silicon quite easily attaches four more electrons. Si atoms have four electrons on the outer s- and p-shells of [Ne] $3s^23p^2$. The other two alloy-forming elements in this system are metals and have radically different electronic configurations: Al–[Ne] $3s^23p^1$ and Y–[Kr] $5d^16s^2$. This reflects the important role of electronic factors in the problem of stability and the formation of intermetallic compounds based on these elements.

Figure 3.30 shows the isothermal sections of the Al–Si–Y ternary system [2, 3], and the analysis shows the possibility of the formation of six three-component compounds in the alloy: $\tau_1(Y_5Al_{14}Si)$, τ_2 (Y_6Al_3Si), $\tau_3(Y_{33.3}Al_{46.7}Si_{20})$, $\tau_4(YAl_2Si)$, $\tau_5(Y_{33.3}Al_{36.7}Si_{30})$, $\tau_6(YAl_2Si_2)$. It is obvious that the presence of oxygen atoms in the surface layer of the alloy will introduce corrections into the phase composition of the material.

Figure 3.31 shows the results of the analysis of the structure and phase composition of layers 1 and 2. It is clearly seen that these layers have a columnar structure. The transverse dimensions of the columns of layer 1 vary from 60 nm to 75 nm. The transverse dimensions of the columns of layer 2 vary from 250 nm to 600 nm. Indication of the microelectron diffraction pattern (Fig. 3.31 b) indicates that these layers are formed by yttrium oxide Y_2O_3.

Layer 1 and layer 2 contain inclusions of the second phase. Particles are rounded; particle sizes range from 5 nm to 12 nm (Fig. 3.31 d). The microelectron diffraction pattern shown in Fig. 3.31 b indicates that these particles are aluminosilicates of the composition Al_2SiO_5.

The results of electron microscopic microdiffraction analysis of the structural phase state of layer No. 3 are presented in Fig. 3.32. It is clearly seen that the layer has a nanocrystalline structure, the crystallite sizes vary within (5–10) nm (Fig. 3.32 a). The crystallites are combined in misoriented regions, the sizes of which are (80–150) nm (Fig. 3.32 c). Indication of the microelectron diffraction pattern obtained from such a structure (Fig. 3.32 b) made it possible to extract reflections of alumina of the composition Al_2O_3 (Fig. 3.32 c) and aluminosilicate of the composition Al_2SiO_5 (Fig. 3.32 d).

Figure 3.33 presents the results of electron microscopic microdiffraction analysis of the structure and phase composition of layer No. 4 indicated in Fig. 3.28. This layer, like layer No. 3, has a nanocrystalline structure. The crystallite sizes vary up to 50 nm. Microdiffraction analysis of layer No. 4 followed by microelectron diffraction patterns revealed reflections of yttrium aluminide of the composition Y_3Al_2 (Fig. 3.33 c) and yttrium silicate of the composition $Y_2Si_2O_7$ (Fig. 3.33 d).

Thus, liquid-phase alloying of an Al–Si alloy with yttrium and oxygen atoms, which is realized under the conditions of ultrahigh heating and cooling rates that occur under the influence of PMPJ, allows one to form a multilayer multiphase nanocrystalline structure in the surface layer of the material, which is mainly represented by oxides and silicates of aluminium and yttrium.

A thermodynamic analysis of phase transformations taking place under equilibrium conditions in the Al–Si–Y system has been performed; the possibility of the formation of six ternary compounds in the alloy is revealed: $\tau_1(Y_5Al_{14}Si)$, $\tau_2(Y_6Al_3Si)$, $\tau_3(Y_{33.3}Al_{46.7}Si_{20})$, $\tau_4(YAl_2Si)$, $\tau_5(Y_{33.3}Al_{36.7}Si_{30})$, $\tau_6(YAl_2Si_2)$. The surface layer of the alloy was modified by plasma treatment, which forms during the electric explosion of aluminium foil with a weighed portion of yttrium oxide powder. Using electron diffraction microscopy, an analysis was made of the elemental and phase composition and defective substructure of the surface modified layer. The formation of a multilayer, multiphase nanocrystalline structure, mainly formed by oxides and silicates of aluminium and yttrium, has been revealed.

3.5. Modelling of processes under the influence of a pulsed multiphase plasma jet

One of the promising pulsed methods is the method of impacting materials with products of electrical explosion of conductors with coaxially arranged electrodes. The products of the electric explosion of the conductor contain solid particles and a gas-plasma component. Powder is placed in the region of the explosion, and a guide channel is installed above the electrodes. Due to high pressures, powder particles have velocities of the order of 1 km/s. The resulting heterogeneous flow acts on the surface of the workpiece. This processing method allows you to add various powders to the surface, and thereby create the appropriate modification of the sample. Under the influence of PMPJ, the sample material melts to a certain depth, and therefore the condensed component penetrates into the surface layers, thereby doping [4].

The method of exposure of the surface to PMPJ includes the following processes:

- electric explosion (electric discharge and the formation of metal plasma, plasma expansion under the influence of thermal and magnetic pressures, acceleration of the plasma powder);

- formation of a heterogeneous plasma flow in the channel;

- interaction of a heterogeneous flow with a surface;

- introduction of accelerated powder particles into the surface layer;

- thermal effect on the sample material;

- evaporation and melting of the material of the surface layer;

- hydrodynamic flow in the molten layer;

- solidification of the molten alloy;

- structural - phase transformations in the heat-affect zone.

Of the presented physical processes, processes leading to the creation of an intermittent effect on the target and the formation of micro- and nanorelief are of the most importance. This chapter presents the mathematical and numerical models of these processes.

3.5.1. Mathematical model for the formation of a heterogeneous plasma flow

The processes in plasma accelerators are characterized by a complex set of hydrodynamic, thermal, and electromagnetic phenomena; therefore, for mathematical modeling it is necessary to use the system of equations of unsteady magnetogas dynamics of a radiating plasma. Due to the complexity of the calculations, mathematical modeling was carried out using an independently developed computer calculation program for the processes of formation and interaction of heterogeneous plasma flows with the surface of metals and alloys. The solution of these equations, because of its complexity, is possible only by numerical methods that would allow us to describe in detail for the end plasma accelerator the interaction mechanisms of the plasma focus (zone I) with a stream flowing perpendicular to the axis (zone III) (Fig. 3.34).

The presence of a free surface dramatically complicates the task even for numerical simulation of the gas-dynamic problem, and therefore the formulation and numerical solution of the magnetoplasma problem is the subject of a special study. On the other hand, the well-known numerical models of electromagnetic accelerators contain various serious assumptions that can significantly distort the integral parameters of plasma flows. Therefore, when modelling plasma flows, certain methods and models have been developed related to the simplification of the initial system of equations and allowing one to relate the installation parameters to the parameters of plasma flows. Such models for a coaxial accelerator ('washer model') and for a gas-injected accelerator ('snow thrower model') were proposed in the 60s, with the help of which a number of optimization studies were carried out. For the end accelerator, considered here a simplified model was not previously proposed.

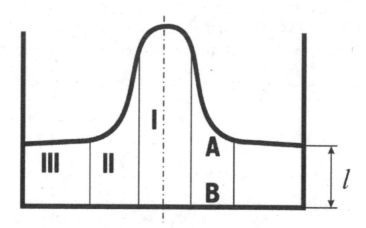

Fig. 3.34. Exploded strip expansion diagram: I – axial current region, II – transition region (current reversal), III – radial spreading region.

Below we consider a simplified problem of flow and discharge motion for an end accelerator. The study of the dynamics of the zone outside the electrodes (zone III Fig. 3.34) is suitable for this role. As a model, we choose the movement of the plasma washer ring, the inner diameter of which commutes with the jet, and the outer one with the peripheral electrode of the unit (Fig. 3.35). The task is to calculate the parameters of a plasma jumper moving between coaxial electrodes under the action of its own magnetic field and gas-dynamic pressure. Below we obtain the equations of motion and the Kirchhoff equation, and we choose the initial conditions taking into account the specifics of our situation.

A diagram of the plasma washer expansion is shown in Fig. 3.35, which indicates the separation of the battery current in two directions along the washer, which closes on the battery and along the axis that flies out of the accelerator. Then the current strength of the battery can be represented $I_0 = I_1 + I_2$, where I_1, I_2 are the currents, respectively, along the washer and along the axis.

This assumption is based on the explanation of the difference in the amplitude values of the current through the exploding foil and the deformable plate. Therefore, this assumption can be used in constructing an approximate mathematical model. For simplicity, we take a triangular velocity profile, the pressure at $z = 0$ depends on the average values of temperature and density, and the resistance to motion is proportional to the square of the speed.

As the initial equations of the model, we choose the laws of change in momentum and mass and the Kirchhoff law:

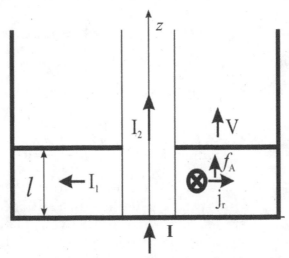

Fig. 3.35. Scheme of the process of expansion of a plasma stream.

$$\frac{\partial \rho}{\partial t}=-\frac{\partial \rho V}{\partial z}, \frac{\partial \rho V}{\partial t}=-\frac{\partial \rho V^2}{\partial z}-\frac{\partial p}{\partial z}+f_A,$$

$$Q/C_0 +\frac{d}{dt}(L_0 I +\Phi)+(R_0 I +R_p I_1 +U_E)=0. \tag{1}$$

where V is the velocity along the z axis, ρ is the plasma density, p is the pressure, f_A is the volume density of the Ampere force, Q is the charge on the batteries during discharge, I is the current through the battery, I_1 is the current through the plasma washer and the parameters of the problem: C_0, L_0, R_0 is the battery capacity, inductance and battery resistance, Φ, R_p is the magnetic flux of the plasma washer and its resistance, U_0, U_E is the voltage on the batteries, plasma-induced voltage. Let us determine the integral characteristics of the flow: momentum, washer mass, Ampere force, magnetic flux of the plasma washer and its resistance, the voltage induced in the plasma due to its motion by the formulas:

$$P=\int_0^l \int_a^b \rho V(z,r,t)2\pi r\, drdz ; \quad M=\int_0^l \int_a^b \rho(z,r,t)2\pi rdrdz$$

$$F_A=\int_0^l \int_a^b j_r B_\phi(z,r,t)2\pi rdrdz ; \quad U_E=\int_0^l \int_a^b VB_\phi(z,r,t)2\pi rdrdz \tag{2}$$

$$\ddot{O}=\int_0^l \int_a^b B_\phi(z,r,t)2\pi rdrdz ; \quad R=(\int_0^l \int_a^b (j_r(z,r,t)2\pi/\sigma)2\pi rdrdz)/I$$

Given that $\dfrac{d}{dt}\displaystyle\int_0^l f(z,t)dz=\dot{i}\cdot f(l,t)+\int_0^l \dfrac{\partial}{\partial t}f(z,t)dz$ provided: $f(0, t) = 0$ and $\dot{l}= V$. Averaging the first two equations (1) over space, we obtain the equations of motion and mass change for the entire washer as a whole:

$$\frac{dP}{dt}=F_A -\rho_0 V^2 S +RT\frac{M}{l\mu}, \quad \frac{dM}{dt}=0 \tag{3}$$

When deriving equations (6), we used: $V(0, t) = 0$, the ideal gas equation $p(0, t) = RT\rho/m$, and the law of resistance in the form $p(l, t) = \rho_0 V^2$. Here, ρ_0, T, m, R is, respectively, the density of the environment, temperature and molar mass of the plasma, and the universal gas constant. The expressions for pressure reflect the

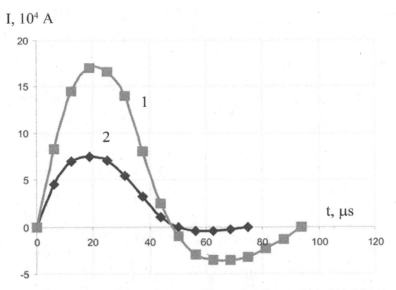

Fig. 3.36. Current strength vs. time (oscillographs (bandwidth 700 MHz)). 1 – shortcut 2 – spark discharge through plasma.

nature of the resistance to the motion of the plasma bunch and the gas-dynamic pressure due to the high temperature. From the second equation (3) it follows that the mass of the washer remains constant. Assuming that the velocity distribution along the (z) axis is triangular with a maximum V at $z = l$, then from the definition of the momentum P from (2) it follows $P = MV/2$.

We express the Ampere force F_A through the current in the circuit. We believe that the currents in zones I and III are proportional to the battery current $I_1 = kI$, $I_2 = (1 - k)I$, and they are homogeneous, then taking into account that $B_\varphi = \mu_0 \cdot I_2/2\pi \cdot r$ and $j_r = I_1/2\pi \cdot l$ we obtain $F_A = I_1 I_2 \cdot A$, where $A = \mu_0 \cdot \ln(a/b)$ where a and b are the geometric dimensions of the peripheral and central electrodes, respectively. Introducing the notation $A_1 = 2k(1-k)A/M$, $A_2 = RT/\mu$, $\alpha = 2\rho_0 S/M$, the system of equations (3) can be reduced to the system:

$$\frac{dV}{dt} = A_1 \cdot I^2 + A_2 / l - \alpha \cdot V^2, \quad \frac{dl}{dt} = V. \tag{4}$$

To transform the Kirchhoff equation, we calculate the magnetic flux, plasma resistance, and induced voltage using formulas (5). Magnetic

flux $\Phi = \int_\Sigma Bd\Sigma = \frac{\mu_0 I_1}{2\pi} \left(\int_a^b \frac{l}{r} dr \right) = \frac{\mu_0 I_1}{2\pi} l \cdot \ln(\frac{b}{a}) = A I_1 l.$: From here for the plasma inductance we get $L = \Phi/I_1 = A \times 1$. Induced voltage drop $U_E = VAI_1$. Then the Kirchhoff equation can be written as a system of equations:

$$\frac{d}{dt}(L_0 I + l \cdot AI_1) + (R_0 I + (R_p + VA)I_1) + Q/C = 0$$

$$\frac{dQ}{dt} = I_0 \tag{5}$$

Thus, the derived system of equations of the mathematical model has the form (4) and (5), which must be solved under the initial conditions

$$l(0) = V(0) = I(0) = 0, \ Q(0) = U_0 C_0 \tag{6}$$

Here, U_0 is the initial voltage on the capacitor banks.

To analyze the resulting system of equations, we estimate the quantities included in (4)–(6). For this we use the installation data and material parameters:

- sizes of electrodes: $a = 5$ mm, $b = 25$ mm; $S = 2 \cdot 10^{-3}$ m²;

- inductance and battery capacity: $L_0 = 0.84 \cdot 10^{-7}$ H, $C_0 = 3$ mF;

- characteristic velocity and length of the plasma bunch: $V = 1$ km /s, $l = 50$ mm;

- density of aluminium and air: $\rho = 2.7 \cdot 10^3$ kg / m³, $\rho_0 = 1.3$ kg / m³;

- specific conductivity of the plasma: $\sigma = 3.3 \cdot 10^7$ 1/(Ohm·m);

- thickness and mass of the foil: $h = 10$ μm, $M = 4 \cdot 10^{-5}$ kg;

- battery resistance: $R_0 = 10^{-7}$ Ohms;

- the amplitude value of the current and half-time: $I_t = 10^5$ A, $T_{1/2} = 5 \cdot 10^{-5}$ s.

Then for the system parameters we get: $A = 3.2 \cdot 10^{-7}$ H/m, $A_1 = 6 \cdot 10^{-3}$ H/(m kg), $a = 50$ m⁻¹, $R_V \approx 3.2 \cdot 10^4$ Ohm; $R_p \approx 4 \cdot 10^{-5}$ Ohm. For the first equation of system (4), we obtain estimates of each term $dV/dt \approx V/T_{1/2} = 2 \cdot 10^7$ m/s²; $A_1 I_2$ 12 · 10^7 m/s²; $\alpha V^2 \approx 10 \cdot 10^7$ m/s². For the first equation of system (5), we obtain estimates for each term $U_0 \approx 10^3$ V; $C^{-1} \int_0^t I dt \approx 0.5 I_m T / C = 0.8 \cdot 10^3$ V.

$\frac{d}{dt}(L_0 I) \approx L_0 I_m / T = 0.2 \ 10^3$ V; $d/dt(l \ A)I \approx LAI_m/T = 32.0$ V;
$(R_0 + R_p + VA)I = 32.4$ V.

From the above estimates, it follows that all terms included in the equation of motion (3) are of the same order, and in the system of equations (4) one can leave terms related only to the battery. Then its solution will become independent of the plasma motion and can be represented as: $I = I_m \sin(\omega \times t)$, where $\omega = 2\pi/T$ and $I_m = U_0 C_0 \omega$ are unambiguously expressed in terms of the installation parameters and the initial battery voltage. This representation is not quantitatively consistent with the experimental current data, since the maximum value of the current in the discharge through the foil is 50% of I_m (Fig. 3.36). However, it follows from it that the dependence of the current strength on time in the first half-cycle is close to a sinusoid, but with a different value of Im. Therefore, we consider a simplified model in order to study its ability to simulate the dependence of the coordinate of the third zone on time, provided that the amplitude value of the current is a fitting parameter.

We study the possibility of mathematical modelling of the motion of the third zone only taking into account the magnetodynamic pressure, i.e. at $A_2 = 0$. Then (2) takes the form:

$$\frac{dV}{dt} = A_1 I_m^2 \sin^2(\omega t) - \alpha V^2, \dot{l} = V \qquad (7)$$

With initial conditions: $V(0) = l(0) = 0$. We introduce dimensionless variables:

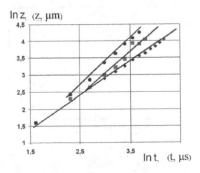

Fig. 3.37. Simulated movement of plasma fluxes vs. experimental data. The points designate experimental data at ♦ – U_0 = 1.0 kV; ■ – U_0 = 1.2 kV; ● – U_0 = 1.6 kV.

$$x = \omega t, \ y = V / V_0, \ z = l / l_0 \qquad (8)$$

then we write the system (7):

$$y' + y^2 = 2f \sin^2(x), \ z' = y; \quad y(0) = z(0) = 0 \qquad (9)$$

where $V_0 = \omega/\alpha$, $l_0 = V_0/\omega$, $f = A_1 l^2 m\alpha/2\omega$.

In the system of equations (9), a dimensionless quantity f is introduced, on which, as a parameter, depends the solution of the problem and, therefore, the nature of the development of zone III in time. We are interested in the range of the argument x from zero to π, which corresponds to a change in time from zero to $T_{1/2}$, to which the coordinates of the third zone were measured. Therefore, the mathematical problem is posed as follows: to study the behaviour of the solution to problem (9) depending on the parameter f.

The first equation (9) is the general Riccati equation, which can be reduced to the linear Mathieu equation using the substitution $y = u'/u$. Then (9) transforms to the form $u'' - 2f \sin^2(x)u = 0$, $u(0) = 1$, $u'(0) = 0$ and subsequent substitution gives:

$$2\xi(1-\xi)\eta'' + (1-2\xi)\eta' - f\xi\eta = 0, \ \eta(0) = 1, \eta'(0) = 0 \qquad (10)$$

This is a linear differential equation of the second order with variable coefficients, the solution of which is in the form of a series

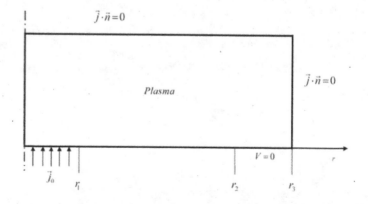

Fig. 3.38. Computational scheme of the magnetohydrodynamic model (j_0 – initial electric current, V – electric potential, r_1 – radius of central electrode, r_2 – internal radius of peripheral electrode, r_3 – external radius of peripheral electrode).

$\eta(\xi) = \sum\limits_{k=0}^{\infty} c_k \xi^k$. After the transformations, we obtain the recurrence relation:

$$c_k = (2c_{k-1}(k-1)^2 + fc_{k-2})/(k(2k-1)), \quad c_0 = 1, \quad c_1 = 0 \qquad (11)$$

Here, we limited ourselves to thirty terms in ξ, while the parameter f remains undefined. Next, we construct the theoretical dependence of ln (z) on ln (x) and compare it with the experimental dependence of ln (z) on ln (t). If the experimental dependence of ln (z) on ln (t) is close to linear, then the theoretical dependence for any values of the parameter f is not current. This indicates the inappropriateness of one-parameter motion modelling, and physically means that only one Ampere force does not provide an acceleration mechanism. We take into account the action of gas-dynamic and magnetic pressures. We turn to dimensionless variables in the system of equations (5) and (6). $x = t/t_0$, $z = l/l_0$, $y = V \cdot t_0 / l_0$, $q = Q/(C_0 \cdot U_0)$. Then we get the system of equations of the model:

$$\frac{dy}{dx} = kf_1 \cdot \left(\frac{dq}{dx}\right)^2 + f_2/z - y^2 \qquad \frac{dz}{dx} = y$$

$$(1 + kf_3 z)\frac{d^2 q}{dx^2} + f_3\left(\frac{dq}{dx}\right)y + q = 0 \qquad (12)$$

$$y(0) = z(0) = \frac{dq}{dx}(0) = 0, \quad q(0) = 1$$

and task parameters:

$$f_1 = 2k(1-k)A(C_0 U_0)^2/(M \cdot l_0),$$
$$f_2 = 2RT(T_{1/2})^2/\mu\, l_0, f_3 = 2k(1-k)A \cdot l_0/L_0 \qquad (13)$$

As follows from the estimates $f_3 \approx 0$. Then only two dimensionless parameters f_1, f_2 appear, which determine the dependence of the coordinate of the third zone on time for different stresses of different values. The system of equations is solved numerically. The obtained system solutions for various parameters f_1, f_2 and their comparison with experimental data are presented in Fig. 3.37.

The theoretical dependence of ln (z) on ln (x) is linear and close to experimental. Thus, the proposed mathematical model makes it possible to adequately describe the experimental data using a

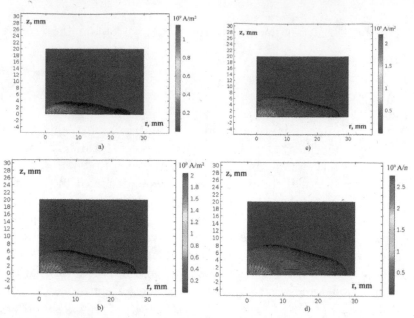

Fig. 3.39. Electric current distribution ($a - 5$ μs, $b - 10$ μs, $c - 11$ μs, $d - 15$ μs).

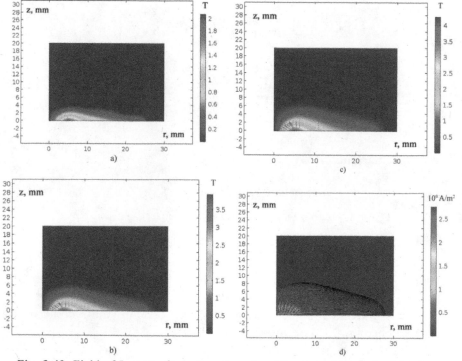

Fig. 3.40. Field of Lorentz force ($a - 5$ μs, $b - 10$ μs, $c - 11$ μs, $d - 15$ μs).

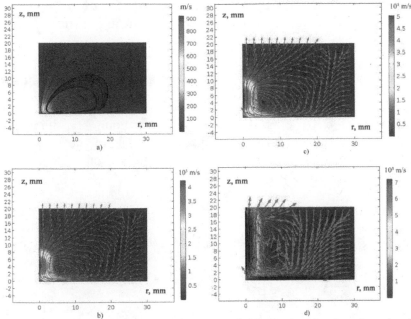

Fig. 3.41. Lines of plasma current and a vector field of velocities (a – 5 µs, b – 10 µs, c – 11 µs, d – 15 µs).

variation of two parameters and indicate the combined action of thermal and gas-dynamic pressures.

Thus, the experimental data obtained in this work on the formation of a heterogeneous plasma flow have been adequately interpreted in the framework of a mathematical model that takes into account the mechanisms of thermal and magnetic pressure.

3.5.2. Numerical model for the formation of heterogeneous plasma flows

The above-proposed model for the formation of a heterogeneous plasma stream gives a qualitative explanation of the behaviour of a heterogeneous plasma stream in the zone where the streamlines are parallel to the dielectric washer. To obtain a more complete picture, as shown in [5–7], it is necessary to resort to numerical magnetohydrodynamic models. An example is the plasma flow model in the Filippov setup [8, 9]. In this work, a single-fluid model of the flow of a current plasma shell was proposed, which showed good agreement with experimental data. In our work, we also apply a single-fluid magnetohydrodynamic model. Figure 3.38 shows the design scheme. In the region $0 < r < r_1$, the input electric current is set, the density of which is equal to $j_0 = \dfrac{I_0}{\pi r_1^2} \sin\left(\dfrac{\pi t}{T}\right)$, where T is the half-period of the pulse, I_0 is the amplitude of the current. In the region $r_2 < r < r_3$, we assume that the potential V is equal to zero.

To calculate the plasma flow at the lower boundary, the condition was set that the velocity was equal to zero, and the pressure was considered equal to zero on the others. The system of equations of the model has the form:

$$\rho\left(\frac{\partial \vec{v}}{\partial t} + \vec{v}\nabla\vec{v}\right) = -\nabla p + \mu\Delta\vec{v} + \vec{F}_V; \nabla \cdot \vec{v} = 0; \nabla \cdot \left(\sigma\nabla V + \sigma\frac{\partial \vec{A}}{\partial t}\right) = 0;$$

$$\sigma\frac{\partial \vec{A}}{\partial t} + \frac{1}{\mu_0}\nabla \times \left(\nabla \times \vec{A}\right) + \sigma\nabla V = 0 \tag{14}$$

where $\vec{v}$ is the velocity vector, p is the pressure, ρ is the density, μ is the dynamic viscosity, $\vec{F}_V$ are the volume forces that are given by the sum of the Lorentz force and gravity $\vec{F} = \vec{j} \times \vec{B} + \rho_0\vec{g}$. The temperature distribution in the calculation domain was determined based on the equation of convective heat conduction with a volumetric heat source.

$$\rho C_p\left(\frac{\partial T}{\partial t} + \vec{v}\nabla T\right) = \nabla \cdot (k\nabla T) + S_V \tag{15}$$

where T is the temperature, C_p is the specific heat, k is the coefficient of thermal conductivity, and S_V are volumetric heat sources. The volumetric heat source is only the Joule effect. The system of equations (14) and (15) was solved by the finite element method. The data for the calculations are given in Table 3.12.

The results of modelling are shown in Figs. 3.39–3.41. Figure 3.39 demonstrates the distribution of the electric current at different moments time. As seen, there are two specific areas there. In the first area the lines of current are perpendicular to the plasma disk, and parallel in the second one, respectively. This fact confirms assumptions about the flow of current when modeling motion of plasma in zone III. Analyzing the fields of Lorentz force (Fig. 3.40), it was revealed that this force would draw a plasma flux in a pitch near the axis perpendicular to the electrode surface.

Figure 3.41 shows the lines of plasma current and a velocity profile. As a matter of fact, there is a jet in the central part of the electrode, and according to high-speed imaging it splits as far as from the surface. In the area around the electrodes (zone III) the plasma flux gets a vortex character. A similar phenomenon is observed in a unit like plasma focus [10,11] and in units with stable plasma pinches [12,13,14,15].

3.6 Conclusions for chapter 3

1. The surface layer of the Al–Si alloy was modified by applying PMPJ to its surface.

2. The optimal processing modes were determined: mode 2 (discharge voltage 2.8 kV; aluminium foil mass 0.0589 g; Y_2O_3 powder mass 0.0589 g.) and mode 5 (discharge voltage 2.6 kV; aluminium foil mass 0.0589 g; mass of Y_2O_3 powder 0.0883 g.)

3. It has been revealed that the impact of PMPJ with optimal parameters leads to the formation of a surface with wear resistance and microhardness, the nanoscale hardness of which is many times higher than the molten state of the alloy.

4. Studies of the structure of the surface profile of the alloy by metallographic and atomic force microscopy showed that PMPJ leads to the formation of a multilayer structure, which consists of a highly porous coating, inhomogeneous in thickness, a layer of liquid-phase alloying and a layer of thermal influence.

5. It was found that PMPJ is accompanied by the formation of a highly porous surface layer with a thickness of 50–80 μm, characterized by heterogeneity in the distribution of alloying elements (silicon, yttrium and oxygen), a submicro- and nanoscale multiphase structure, the strengthening phases of which are particles of silicon, Y_2O_3, YSi_2 and $Y_2Si_2O_7$.

6. The methods of electron diffraction microscopy were used in the analysis of the elemental and phase composition and of the defective substructure of the surface modified layer. The formation of a multilayer, multiphase nanocrystalline structure, mainly formed by oxides and silicates of aluminium and yttrium, has been revealed.

7. It was established that the Al–Si alloy is a multiphase material and contains, in addition to phases based on solid solutions of aluminium and silicon, intermetallic compounds of various compositions. Particles of silicon and intermetallic compounds have a diverse shape (globular, lamellar, needle-shaped or skeletal) and,

Table 3.12. Characteristics of the material and parameters of the unit

Characteristic	Value
density of aluminium and air	2700 kg/m^3
air density	1.3 kg/m^3
plasma conductivity	3.3•107 Sm/m;
foil thickness	10 μm
mass of foil	4•10^{-5} kg
characteristic velocity of plasma clot	1 km/s
characteristic length of plasma clot	50 mm
electrode sizes	a =5 mm, b =25 mm; S = 2•10^{-3} m^2
battery resistance	10^{-7} Ohм
battery inductance	0.84•10^{-7} H
battery capacity	3 mF
current amplitude	10^5 A
half-period time	5· 10^5 s

therefore, can not only strengthen the material, but also be sources of microcracks during the operation of products.

8. A thermodynamic analysis of phase transformations taking place under equilibrium conditions in the Al–Si–Y system has been performed; the possibility of the formation of six ternary compounds in the alloy is revealed: $\tau_1(Y_5Al_{14}Si)$, $\tau_2(Y_6Al_3Si)$, $\tau_3(Y_{33.3}Al_{46.7}Si_{20})$, $\tau_4 (YAl_2Si)$, $\tau_5(Y_{33.3}Al_{36.7}Si_{30})$, $\tau_6(YAl_2Si_2)$.

9. A study of the motion of the plasma flow in zone III using a simplified model showed that it adequately describes the motion of the plasma flow observed in the experiment.

10. Using the finite element method, the plasma current, electric current, and Lorentz force field distributions are obtained.

References for Chapter 3

1. Lyakishev N.P., Diagrams of the state of binary metal systems Moscow, 1996.
2. Geschneidner K. A., Bull. Alloy Phase Diagrams. 1988. No. 9. p. 658–668.
3. Drits M. E., Kuzmina V. I., Tylkina N. I. Russ. Metall. 1980. V. 3. p. 178–181.
4. Bagautdinov A.Ya., et al., Physical fundamentals of electric explosive alloying. Novokuznetsk: SibGIU, 2007.
5. Lerner, M. I. Modern technology for producing nanoscale materials. Tomsk: TPU.
6. Sedoi, V. S., Ivanov Y.F. Nanotechnology. 2008.No. 19. 145710.
7. Bakshi R. B., et al. Zh. Tekh. Fiz. 2013.V. 83. No. 8. p. 4353.
8. Sivkov A.A., et al., Izv. Tomsk. Politekhn. Univ.. 2010.V. 317. No. 4. p. 74–78.
9. Ananiev S.S., et al., Voprosy Atom. Nauki Tekh. 2016.V. 39. No. 2. p. 69–80.

10. Ananyev S.S., Kharrasov A.M., Suslin S.V., J. Phys. Conf. 2016. V. 747. 012014.
11. Volkov N. B., Iskoldsky A. M., J. Phys. Math. Gen. 1993. V.26. p. 6667–6677.
12. Liberman M.A., et al., Physics of High Density Z-pinch Plasmas, Springer, 1999.
13. Auluck S.K.H., Phys. Plasmas 2016. V.23. 122508.
14. Escande D.F., Plasma Phys. Contr. Fusion 2016. V. 58. 113001.
15. Gol'dberg M.M., et al., Sov. Phys. J. 1986. V. 29. Issue 6. p. 428–432.

4

Investigation of the properties, phase composition and defective substructure of the surface layers of Al–Si alloys after the effect of an intense pulsed electron beam

4.1. Determination of changes in microhardness, tribotechnical tests, metallographic analysis of structural changes in Al–Si alloys subjected to electron beam irradiation in various modes

4.1.1 Metallographic analysis of the structure of an Al–Si alloy subjected to electron beam irradiation

Typical images of the structure of the etched thin section obtained by optical microscopy are presented in Fig. 4.1. It can be seen that the material under consideration mainly consists of grains of aluminium and eutectic.

Irradiation of an Al–Si alloy by an electron beam, regardless of the electron beam density, is accompanied by melting and homogenization of the surface layer of the material, which is indicated by lower contrast micrographs in this region and the impossibility of obtaining a clear pronounced structure by chemical etching (Fig. 4.2).

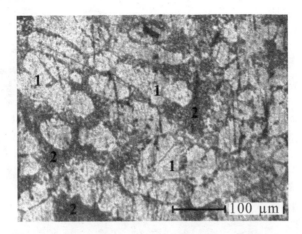

Fig. 4.1. Al–Si alloy structure revealed by optical microscopy (1 – aluminium grains, 2 – eutectic).

It can be seen that the thickness of the layer melted deep into the material increases with increasing electron beam density and amounts to: 24 μm for an energy density of 10 J/cm², 38 μm for 15 J/cm², 42 μm for 20 J/cm², 61 μm for 25 J/cm², 66 μm for 30 J/cm² and 100 μm for 35 J/cm².

4.1.2. Analysis of changes in micro-, nanohardness and plasticity parameter of an Al–Si alloy after the effect of an intense pulsed electron beam

As a characteristic of the mechanical properties of surface layers, we used one of the most accurate and sensitive methods - microhardness measurement. Its differences before and after processing can serve as an indicator of hardening of modified surface layers of metals and alloys.

The microhardness was measured both directly from the side of the subjected modification (Fig. 4.3 *a*), and at different distances from it using a cut thin section (Fig. 4.3 *b*).

The dependence of the change in the microhardness of the surface of the Al–Si alloy on the energy density of the electron beam is shown in Fig. 4.4. An analysis of the graph shows that an increase in the energy density leads to a monotonic increase in the microhardness value on the irradiated surface.

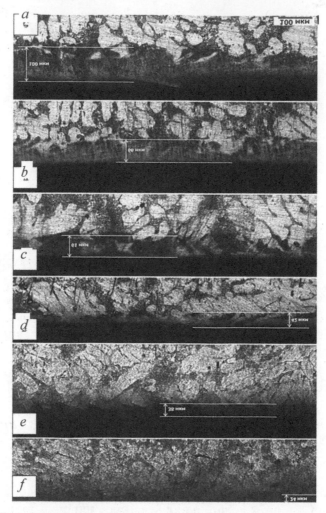

Fig. 4.2. Microphotographs of the cross-sectional structure of the samples subjected to irradiation with an intense pulsed electron beam with different energy density J/cm² (a – 10, b – 15, c – 20, d – 25, e – 30, f – 35).

The maximum increase in microhardness is observed at an electron beam energy density of 30 J/cm². A further increase in the energy density to 35 J/cm² leads to a slight decrease in the microhardness.

Since the maximum values of surface microhardness are observed for values of 25, 30, and 35 J/cm², we studied the distribution of the microhardness profile depending on the distance to the irradiated surface using transverse sections. Since the material contains aluminium grains and a eutectic, microhardness measurements were carried out separately in the grain (Fig. 4.5 c) and in the eutectic (Fig. 4.5 b).

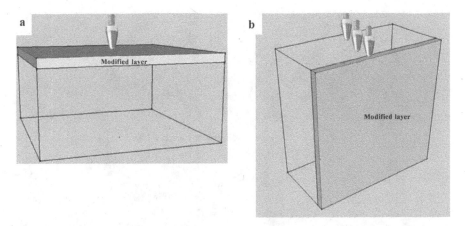

Fig. 4.3. Schematic designation of microhardness measurement sites on samples modified by an electron beam.

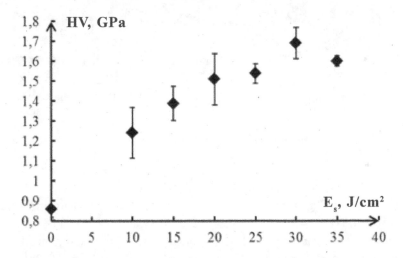

Fig. 4.4. The dependence of the change in microhardness on the energy density of the electron beam directly from the modified side.

It was determined that the microhardness values (Fig. 4.6), both in grains and in the eutectic of modified samples, increase as one approaches the sprayed layer. It was found that, regardless of the treatment regimes, the microhardness of the samples in the irradiated zone is greater than at distances of 90 and 70 μm from the edge of the sample. The analysis of the dependences gives reason to conclude that the microhardness in the eutectic is greater than in grains.

Figure 4.6 *d* shows that the microhardness decreases as it moves into the bulk of the material and at a depth of 90 μm (0.93 ± 0.052

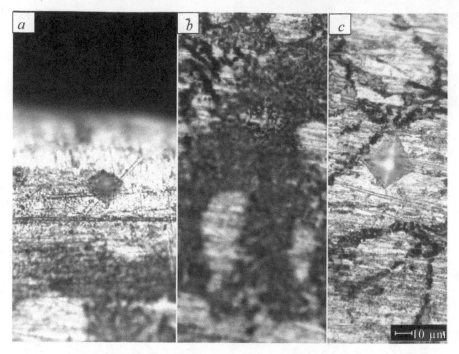

Fig. 4.5. Microstructure of a transverse section of an Al–Si alloy irradiated with an intense pulsed electron beam with indenter prints (optical microscopy); *a* – structure at the irradiated surface; *b* – eutectic at a distance of 70 μm from the surface; *c* – imprint in the grain of aluminium at a distance of 70 μm.

GPa for 25 J/cm², 0.97 ± 0.071 GPa for 30 J/cm², 0.96 ± 0.103 GPa for 35 J/cm²) almost reaches the values of the starting material (0.86 ± 0.04 GPa), regardless of the processing mode. The nature of the dependence also does not depend on the processing mode.

According to the results of studying the effect of IPEB (intense pulsed electron beam) on the microhardness of the surface layers of the material under study, we can conclude that the optimal processing parameters, allowing to almost double the microhardness, are the modes with an energy density of 25, 30, 35 J/cm².

After analyzing the data of changes in the microhardness of the studied samples, it is possible to calculate the plasticity parameter (Fig. 4.7).

It is known that the plasticity characteristic determined by the Vickers method of microindentation has the form: $\delta = 1 - 1.14 \, (1 - \mu - 2\mu^2) \, HV/E$, where HV is the microhardness value, E is the modulus, μ is the Poisson's ratio of the material under study [1].

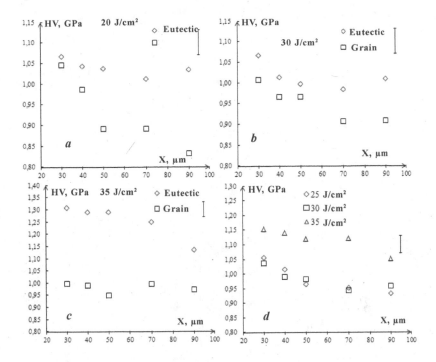

Fig. 4.6. Dependences of the distribution of microhardness values in grains and eutectics at various distances from the modified surface (a – 25 J/cm², b – 30 J/cm², c – 35 J/cm², d – average for the material)

An analysis of the dependences of the plasticity parameter δ of the investigated alloy on the distance to the processing surface shows that in the zone of the modified layer the plasticity parameter has minimal values, regardless of the treatment mode. Moving deep into the material leads to a nonmonotonic increase in the plasticity parameter. It should be noted that the plasticity parameter has the minimum values in the eutectic at the processing mode of 35 J/cm². In addition, the value of the plasticity parameter is greater in grains than in eutectics, irrespective of processing conditions.

A nanohardness profile was constructed using a Nano Hardness Tester NHT-S-AX-000X. The method of measuring nanohardness is based on an analysis of the relationship that varies over time of the load to the penetration depth and indenter footprint (kinematic hardness method).

Figure 4.8 shows images of a thin section surface with indenter prints. It is clearly seen that the dimensions of the prints, and, accordingly, the nanohardness of the analyzed region of the material,

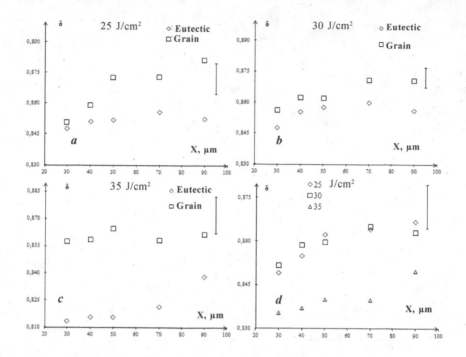

Fig. 4.7. Graphs of the dependence of the plasticity parameter in grains and eutectics at various distances from the modified surface (a – 25 J/cm², b – 30 J/cm², c – 35 J/cm², d – average for the material).

significantly depend on the distance to the irradiated surface (in Fig. 4.8 a, the irradiated surface is indicated by an arrow). Namely, near the irradiated surface, the dimensions of the indenter imprint are noticeably smaller than those located at a large (200 μm, for Fig. 4.8 b) distance from the irradiated surface. Therefore, irradiation of an Al–Si alloy by an electron beam is accompanied by hardening of the surface layer of the sample.

Figure 4.9 shows the results of constructing a nanohardness profile of irradiated samples of an Al–Si alloy. Analyzing the results, it can be noted that the hardness of the studied material varies non-monotonously, reaching maximum values exceeding the initial state hardness by 3.8–4.2 times, in a layer located at a depth of 30–50 μm. In the layer adjacent to the irradiated surface, i.e. located at a depth of ≈5 μm, the value of hardness exceeds the hardness of cast Al–Si alloy (in our case, a layer located at a depth of 500 μm) by ≈1.6 times.

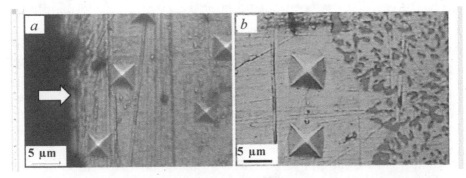

Fig. 4.8. Image of the cross-sectional structure irradiated by an intense pulsed electron beam with indenter prints (SEM); *a* – structure at the irradiated surface; *b* – at a distance of 200 μm from the surface; on (*a*) the arrow indicates the irradiated surface.

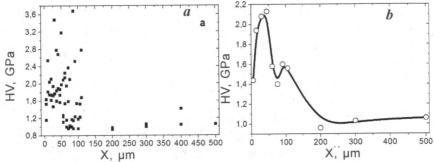

Fig. 4.9. Field of values of nanohardness (*a*) revealed during nanoindentation of a transverse section irradiated by an intense pulsed electron beam; on (*b*) the dependence of the averaged hardness values on the distance from the irradiated surface is given.

4.1.3. Tribological testing of an Al–Si alloy after electron beam irradiation with different energy densities

Tribotechnical tests were carried out according to the pin-on-disc scheme on Oscillating TRIBOtester (TRIBOtechnic) tribometers (with a load of $P = 1$ N, sliding speed of $V = 25$ mm/s) and CSEM CH 2000 (with a load of $P = 2$ N and sliding speed $V = 10$ mm/s; the diameter of the counterbody made of ShKh15 steel was 6 mm) in accordance with ASTM G99. The friction path $S = 20$ m and the radius of the wear path $r = 2$ mm were identical in both cases. The parameters of tribological tests on the Oscillating TRIBOtester tribometer are given in Table 4.1.

The tribological properties of the modified alloy were characterized by a wear coefficient and a friction coefficient

(TRIBOtester device). The installation is based on the principle of a tester, on a pin disk with a rotating disk and a static sample. The main element of the device is a 'pin', having a certain radius and made of a certain material, which is aligned perpendicular to the 'rotating' disk. By changing the load on the pin, it is possible to change the friction coefficient and, therefore, determine what wear occurs between the materials (wear coefficient). The frictiom friction coefficient is measured as the inertial moment arising between the investigated materials.

The load on the pin is provided by a weight that can be moved along the extended bracket. This provides an optimal form of load, and also makes it possible to change the load during testing.

Simultaneously with the increase in microhardness in the irradiated samples, a decrease in the friction coefficient and wear rate is observed (parameter, inverse of wear resistance – see Table 4.2). Compared with the material in the delivery state, with the tri-loading parameters $P = 1$ N, $V = 25$ mm/s, the friction coefficient decreased by ≈ 1.3 times, the wear rate by ≈ 6.6 times. The diagrams characterizing the change in the friction coefficient during tribotechnical tests at a lower load and higher tribo-loading speed are shown in Fig. 4.10.

Comparing the results of tribotesting of the alloy in the cast (Fig. 4.10, curve a) and irradiated (Fig. 4.10, curve b) states, it can be noted that, firstly, the time for the friction coefficient to reach the stationary mode of change in the irradiated sample is significantly longer (not less than 100 s) and, secondly, the amplitude of the friction coefficient of the cast sample is significantly higher

Table 4.1. Conditions for tribological testing of samples on an Oscillating TRIBOtester (TRIBOtechnic)

Parameters of tribotechnical tests		Coating material	
Load	1 N	Material	100Cr6 steel
Sliing speed	25 mm/s	Diameter	6000 mm
Sliding length	20 m	Geometrical form	Sphere
Radius of wear path	2 mm	Young modulus	205000 MPa
Temperature	23°C		
Humidity	50%		
Atmosphere	Air		

($\Delta\mu \geq 0.3$), which may be due to the development of setting processes, chipping of the strengthening particles, as well as the transfer of wear products from the tribocontact zone on the counterbody.

With an increase in the load on the counterbody (tribo-loading parameters $P = 2$ N, $V = 10$ mm/s) for the irradiated sample (at $W = 35$ J/cm^2), the friction coefficient relative to the sample in the delivery state remained practically unchanged (cf. $\mu = 0.42$ relative to 0.45, see Table 4.2, Fig. 4.10). At the same time, in comparison with the conditions of applying a lower load / higher tribo-loading speed, the measured value of the friction coefficient of the sample without irradiation decreased to a greater extent from $\mu = 0.56$ to 0.47. The reasons for this will be discussed below.

In tribological tests at a higher speed/lower load, the wear rate of an non-irradiated sample is approximately 2 times higher than at $P = 2$ N, $V = 10$ mm/s (cf. $I = 4.9 \cdot 10^{-3}$ relative to $2.7 \cdot 10^{-3}$ mm^3/N·m). In the case of a sample with a modified surface, this parameter decreases by several times and practically does not depend on the parameters of tribo-loading (cf. $I = 0.74 \cdot 10^{-3}$ relative to $0.78 \cdot 10^{-3}$ mm^3/N·m). The reason for this, in our opinion, should be the development of the processes of setting and chipping of hardening particles, characteristic of aluminium alloys, and its suppression due to the modification of the structure of the surface layer during irradiation.

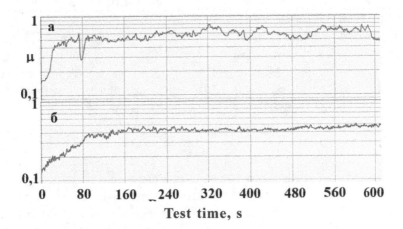

Fig. 4.10. Dependence of the friction coefficient m on the time of tribotechnical testing of the material in the cast state (a) and after irradiation with an intense pulsed electron beam (b); tribo-loading parameters: $P = 1$ N, $V = 25$ mm/s.

Table 4.2. The results of measuring the tribological properties of a material irradiated by an intense pulsed electron beam

State	Friction coefficient		Rate of wear, i, 10^{-3}. mm²/N m	
Sliding speed/ load	$P = 1$ N $V=25$ m/s	$P = 2$ N $V =10$ mm/s	$P=1$ N $V = 25$ mm/s	$P = 2$N $V = 10$ mm/s
Cast state	0.56	0.47	4.9	2.7
State after irradiation, $W=35$ J/cm²	0.42	0.45	0.74	0.78
State after irradiation $W=15$ J/cm²		0.37		0.93

The curves characterizing the change in the friction coefficient during tribotechnical tests at the parameters of tribo-loading $P = 2$ N, $V = 10$ mm/s are shown in Fig. 4.11. It can be seen that, in contrast to the tribo-loading conditions described above ($P = 1$ N, $V = 25$ mm/s), the running-in time for the sample in the delivery state (the friction coefficient value reaches the steady-state change mode) is significantly longer (in terms of the friction path, it is not less than 0.003 km). On the other hand, the value of the friction coefficient for the samples in the delivery state and after irradiation (at $W = 35$ J/cm²) is almost identical. In our opinion, the predominant reason for this is a decrease in the sliding speed, which, due to a decrease in frictional heating, should reduce the intensity of the development of the processes of setting and chipping of the strengthening particles.

Here, for comparison, a graph of the friction coefficient of a material irradiated with a lower dose ($W = 35$ J/cm²) is shown, for which it can be seen that a decrease in m corresponds to a decrease in wear rate compared to unirradiated material (cf. $I = 0.93 \cdot 10^{-3}$ mm³/N·m with respect to $2.7 \cdot 10^{-3}$ mm³/N·m, Table 4.2).

Thus, an increase in the sliding friction velocity from $V = 10$ to 25 mm/s, with a half load on the counterbody, in the case of an unirradiated sample, is accompanied by a twofold increase in the wear rate, while this parameter does not change for an irradiated sample. This is evidence of the effectiveness of the proposed method for increasing the tribotechnical properties of the Al–Si alloy by IPEB.

Additional information on the causes of the observed patterns was obtained from the analysis of tribotrack profiles of the tested samples. In Fig. 4.12 such data are given for the conditions of tri-

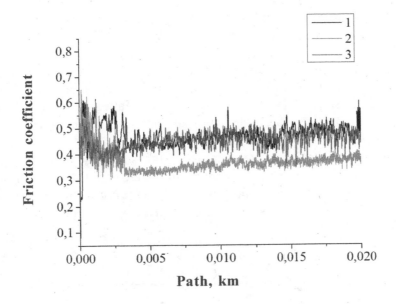

Fig. 4.11. Dependence of the friction coefficient m on the time of tribological testing of specimens in the molded state (*a*) and after irradiation with an intense pulsed electron beam (*b*) at the triboload parameters P = 2 N, V

loading $P = 1$ N, $V = 25$ mm/s. It can be seen (Fig. 4.12 *a*) that the profile of the friction track is extremely rugged, which, most likely, is associated with the development of setting and chipping processes, accompanied by tearing of the material from the friction surface. The quantitative parameters of the tribotrack are shown in Table 4.3. It can be seen that in the initial state of the material with a hole width of about 400 µm, the maximum depth reaches almost 26 µm; in this case, the area of the hole (in the local section) can be estimated as 6.2 µm². At the same time, in the irradiated sample, the tribotrack width is almost the same (400 µm), however, the depth of the hole » 7.2 µm and its area 1.3 µm² decreased significantly. Again, it should be noted that for the used parameters of the tribo-loading, the profile of the wear well also looks very heterogeneous and non-smooth / rugged (Fig. 4.12 *b*).

A twofold increase in load to $P = 2$ N, with a decrease in speed to $V = 10$ mm/s, resulted in a smoother wear pattern, as evidenced by the corresponding tribreck profiles (Fig. 4.13). It can be seen that the profile of the wear hole for both wear conditions has become more oval and has a more even outline. For a non-irradiated sample, the maximum depth of the well does not exceed 21 µm with its full width of the order of 700 μm, which is higher than with a smaller load on the counterbody (Fig. 4.12 *a*). In the case of an irradiated sample (W = 35 J/cm²), the tribotrack width decreased to 500 μm, and its depth does not exceed 12.1 μm (Fig. 4.13 *b*). In this case, the change in the area of the well is smaller than it was at lower tribo load: cf. 3.4 times relative to 4.9 times (see Table 4.3). For a lower dose of IPEB (W = 15 J/cm²), the parameters of tribotrack are not much different (see Table 4.3).

Additionally, studies were made of the surfaces of the friction tracks. It can be seen that the friction surface of the unirradiated sample looks less uniform, and a large number of dark areas testifies to the development of setting and chipping processes. These

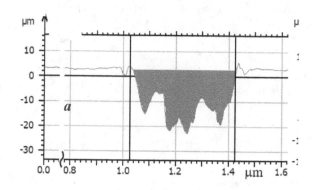

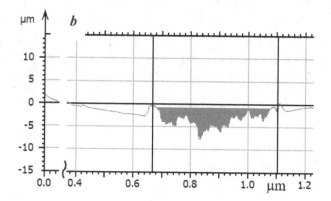

Fig. 4.12. The profiles of the track of wear of the samples in the cast state (*a*) and after irradiation with an intense pulsed electron beam (*b*), with the parameters of tribo-loading P = 1 N, V = 25 mm/s.

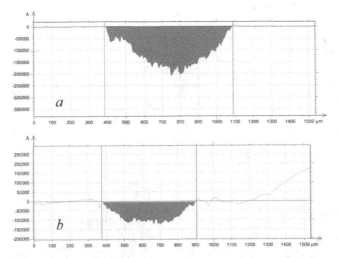

Fig. 4.13. The profiles of the wear track of the samples in the cast state (*a*) and after irradiation with an intense pulsed electron beam (*b*). with the parameters of the tribo-load $P = 2$ N, $V = 10$ mm/s.

phenomena are manifested to a much lesser extent in the irradiated sample. Analysis of the counterbody surfaces shows that during tribo-tests, processes of transfer of wear products develop; the counterbody for a non-irradiated sample contains a large number of dark areas, which are just the result of adhesion of wear products (Fig. 4.14 *b*). On the surface of the counterbody for the irradiated material, this process develops and manifests itself to a much lesser extent. In general, according to the analysis of profiles of wear tracks, it follows that during tribotechnical tests of cast material wear tracks are formed, which are significantly larger and have a large difference in track depth.

Table 4.3. Quantitative characteristics of the wear track of samples in cast and irradiated states

Condition	Maximum depth, μm		Well area, μm²	
Sliding speed/load	$P = 3$ N $V = 25$ mm/s	$P = 2$ N $V = 10$ mm/s	$P = 1$ N $V = 25$ mm/s	$P = 2$ N $V = 10$ mm/s
Cast	25.9	21.0	6.17	8.52
After irradiation, W = 35 J/cm²	7.2	12.1	1.27	2.47
After irradiation, W = 15 J/cm²		14.3		2.95

Fig. 4.14. Optical images of friction paths (*a*) and counterbody (*b*) after the end of tribo-tests with the tribo-loading parameters $P = 2$ N, $V = 10$ mm/s for samples in the cast state and after irradiation with an intense pulsed electron beam.

It is obvious that the significant increase in microhardness and tribotechnical properties of the Al–Si alloy shown above is due to the modification of the elemental and phase composition, as well as to the state of the defective substructure of the surface layer of the material initiated by high-speed heat treatment that occurs when the material is irradiated with an intense pulsed electron beam [2–6].

4.1.4. Comparative analysis of changes in the strength properties of the surface layers of an Al–Si alloy subjected to processing by an electron beam and a multiphase plasma jet of the Al–Y₂O₃ system

A comparative analysis of the results of studies of an alloy of the Al–Si system subjected to PMPJ with yttrium oxide powder Y_2O_3 and electron-beam processing was carried out. In the course of the work, it was found that PMPJ with yttrium oxide Y_2O_3 powder leads to the formation of a composite coating, the coating is characterized by heterogeneity in thickness, high porosity and a large number of microcracks. Irradiation of this alloy with a high-pulse electron beam leads to the formation of a molten modified homogeneous layer with a changed structure. The thickness of the melted layer depends on the processing parameters. A comparison of the results of mechanical tests shows that the microhardness values, both in grains and in the eutectic of modified samples, increase as one approaches the sprayed layer or in the melted layer, in the case of IPEB. It was found that, regardless of the processing regimes, the microhardness of the samples in the zone of the modified modification is greater than at distances of 90 and 70 μm from the edge of the sample. So, for the PMPJ, the optimal regimes were 2 ($m_{Y_2O_3}$ = 0.0589, U = 2.8 kV) and 5 ($m_{Y_2O_3}$ = 0.0883, U = 2.6 kV); under these processing conditions, the microhardness of the surface layers was 1.3 GPa in mode 2; 1.6 GPa in mode 5. Evaluation of this parameter with IPEB shows that the hardness of surface layers close to the irradiated surface varies from 1.05 to 1.16 GPa, depending on the energy density of a bunch of electrons. The analysis of the dependences gives the basis to conclude that the microhardness of the Al–Si alloy in the eutectic is greater than in grains, regardless of processing. An analysis of the dependences of the ductility parameter δ of the alloy on the distance to the processing surface shows that in the zone of the modified layer and coating, the ductility parameter has minimal values, regardless of the treatment mode. Analyzing the results, it can be noted that the nanohardness of the material treated with an intense electron beam varies non-monotonously, reaching maximum values exceeding the hardness of the starting material by 3.8–4.2 times. The nanohardness of the coating obtained as a result of PMPJ also varies non-monotonically and depends on the porosity of the coating; as it moves toward the porous coating layer, the nanohardness decreases by 24% relative to the dense layer. The hardness of the coating reaches 2.20 GPa. An analysis of the results shows that the wear resistance of the investigated material after PMPJ increased, compared with the wear resistance of the initial state, by more than 28 times; the friction coefficient decreased by more than 2 times. Tribotechnical tests of the Al–Si alloy irradiated with an intense electron beam show that, in comparison with the material in the initial state, the friction coefficient decreased by ≈1.3 times, the wear rate by ≈6.6 times.

A comparative analysis of the data obtained as a result of the work on the influence of PMPJ and intense electron beams allows us to determine the vector of further research, the results of which are presented in the following chapters of this book. Since the formation of a coating and a remelted layer occurs during PMPJ and IPEB, respectively, the mechanical characteristics of which are many times greater than the characteristics of the alloy in the cast state, it becomes relevant to conduct studies on the combined surface treatment of Al–Si alloy. The combined treatment will consist of coating the Al–Ti–Y$_2$O$_3$ system on an Al–Si alloy substrate, followed by processing by intense electron beams, which, in our opinion, will lead to the formation of a uniform coating, characterized by higher values of mechanical characteristics compared to individual types of treatments.

4.2. Atomic force microscopy of samples of Al–Si alloys exposed to an electron beam of submillisecond duration

4.2.1. Results of atomic force microscopy of samples exposed to an intense pulsed electron beam

The use of atomic force microscopy to solve metallographic problems allows obtaining information on the surface topography in high resolution, as well as its local properties [7]. It was shown in [8–10] that atomic force microscopy methods can be successfully applied to study the structure of 40Kh13 steel after tempering and hardening, as well as for certification of micro- and nanostructures of Kh18 stainless steel. So, it was found that, depending on the hardness, the structural components of the metallographic thin section of Kh18 steel are etched differently and are located at different heights relative to each other [10]. The semi-contact atomic force microscopy method used in [11] made it possible to study the structure of structural steel at the nanoscale and to establish that the effect of high temperatures on the material thickens the grain boundary for a long time due to an increase in the carbon content at the boundary, leading to its carbidization. In [12], nanodomain structures arising in lithium niobate single crystals as a result of intense pulsed laser radiation were studied. Atomic force microscopy made it possible to establish that the growth of domain rays begins with the formation of chains consisting of isolated nanodomains with a width of 30 nm.

In [13], samples of technically pure titanium used for dental implants were treated with an electron beam in various modes and studied using atomic force microscopy. The roughness Ra of the samples processed in three modes was determined, which differ in the width of the groove left from scanning by the electron beam and the initial sample with a mirror surface. A surface with a depth of 30 μm was treated with an electron beam, which led to nonequilibrium crystallization with the formation of martensitic and lamellar phases. The results showed that the roughness of the initial sample is 7 nm, and the roughness of the processed samples lies in the range from 20 to 40 nm.

Thus, the above results of the use of atomic force microscopy in studying the structure and properties of various metals, in particular, processed by an electron beam, indicate the advisability of using this method in the present work. This will expand the amount of information obtained, the analysis of which will make it possible to control the quality of the surface layers of aluminium alloy samples processed by a pulsed electron beam at a new level.

In the present work, samples treated with an electron beam with an electron beam energy density ranging from 10 to 35 J/cm² were studied using atomic force microscopy. The remaining parameters remained the same for all modes: accelerated electron energy 17 keV, pulse duration of the electron beam 150 μs, number of pulses 3, pulse repetition rate 0.3 s⁻¹, residual gas pressure (argon) in the working chamber of the apparatus $2 \cdot 10^{-2}$ Pa.

Topographic images of the transition zone between the coating and the substrate were obtained on a direct section in the tapping mode of an atomic force microscope. The average roughness of the coating and substrate was determined for samples processed in various modes. The average roughness values Ra were obtained from three images for each mode. The roughness of the substrate was determined at a distance of ≈30 μm from the boundary of the treated layer.

The profile image of the initial sample obtained using atomic force microscopy is shown in Fig. 4.15.

The image of the initial sample is characterized by the presence of a dendritic grain structure and intergranular eutectic. The grain boundaries have inclusions of intermetallic compounds mainly consisting of copper, manganese and nickel, which was recorded by scanning electron microscopy. The roughness of the initial material sample is ≈60 nm.

Images of profiles of samples processed in different modes are presented in Figs. 4.16–4.21.

A typical image of a sample treated according to mode No. 1 obtained using atomic force microscopy is shown in Fig. 4.16. The image was obtained on a transverse section near the edge of the sample. Due to the low energy density of the electron beam, the structure of the treated layer does not undergo significant changes. However, in the near-surface layer, an accumulation of small precipitates of the second phase from 2 to 15 μm in size was found at a distance of 10–15 μm from the edge. The roughness of the processing layer Ra is 99 nm, that of the substrate is 77 nm.

Figure 4.17 shows the structure of a sample treated with an electron beam with an electron beam energy density of 15 J/cm². The eutectic is a mixture of components that have sizes from 1 to 3 μm. It is seen that the surface layer has a fine-grained structure, its thickness is 30 μm. The roughness of the cross section of the treated layer and the substrate, Ra is the same and also is 30 nm.

The cross-sectional structure of a sample treated with an electron beam with an electron beam energy density of 20 J/cm² is shown in Fig. 4.18. The image shows a fine-grained cellular structure, as well as an eutectic and partially remelted grain boundary (indicated by numbers and an arrow in Fig. 4.18). A study using atomic force microscopy confirms the absence of intermetallic compounds in the structure of the treated layer. Inclusions of the second phases in the eutectic have average sizes from 1 to 5 μm. The roughness of the treated layer Ra is equal to 27 nm, that of the substrate is 57 nm.

Presented in Fig. 4.19, the cross-sectional structure of a sample treated with an electron beam with an electron beam energy density of 25 J/cm² has a fine-grained treated layer and grains with grain boundaries (indicated by numbers in Fig. 4.19). The roughness of the

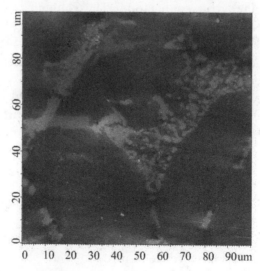

Fig. 4.15. Atomic force microscopy of the initial state of samples. 1 – intergranular boundary, 2 – dendritic grain, 3 – intergranular eutectic.

treated layer *Ra* is equal to 27 nm, the substrate is 45 nm. Also in the structure of the processing layer are particles that were inherited from grain boundaries and were not recrystallized (indicated by white arrows in Fig. 4.19)

Figure 4.20 shows a cross-sectional image of an Al–Si alloy sample treated with a high-intensity pulsed electron beam of submillisecond duration of exposure, the electron beam energy density of 30 J/cm². In the treated layer there are directionally

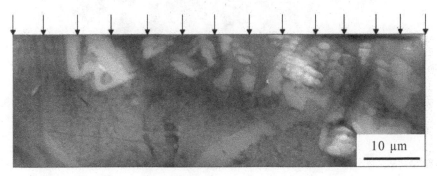

Fig. 4.16. Atomic force microscopy of a sample treated with a high-intensity pulsed electron beam of submillisecond exposure time, electron beam energy of 10 J/cm² (arrows indicate the direction of exposure to the electron beam)

recrystallized grains. The number 2 indicates the incompletely fused grain boundary. The roughness of the treated layer, *Ra,* is 33 nm, the substrate is 51 nm.

Figure 4.21 shows a cross-sectional image of a sample treated with an electron beam with an electron beam energy density of 35 J/cm². The structure is a treated layer 1, an intergranular boundary 2, and a grain body 3 and 4. The roughness of the treated layer *Ra* is 17 nm, and the substrate is 45 nm.

Table 4.4 shows the roughness values of the coatings and substrates of the samples subjected to IPEB under various conditions. The average roughness *Ra* of the treated layer varies from 17 to 99 nm, depending on the energy density of the electron beam. The average roughness *Ra* of the substrate is in the range from 30 to 77 nm. The average value of the coating roughness, *Ra*, equal to 99 nm, was obtained in the sample, processed by an electron beam with an electron beam energy density of 10 J/cm², the smallest 17 nm, with an energy density of 35 J/cm².

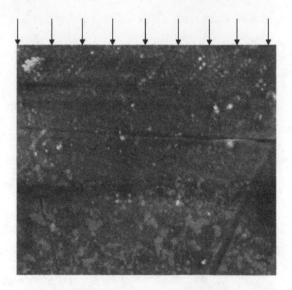

Fig. 4.17. Atomic force microscopy of a material treated with a high-intensity pulsed electron beam of submillisecond exposure time, electron beam energy of 15 J/cm^2 (arrows indicate the direction of exposure to the electron beam).

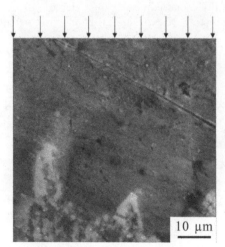

1 - region of fine-grained cellular structure, 2 - eutectic structure of the substrate, the arrow indicates a partially remelted grain boundary

Fig. 4.18. Atomic force microscopy of a sample treated with a high-intensity pulsed electron beam of submillisecond duration of exposure, electron beam energy of 20 J/cm^2 (arrows indicate the direction of exposure to the electron beam).

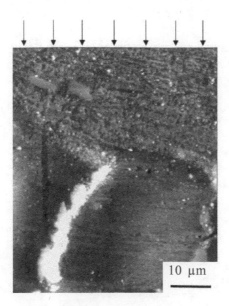

1 – processed layer, 2 – grain body, 3 – grain boundary, arrows indicate particles inherited from the grain boundary

Fig. 4.19. Atomic force microscopy of a sample processed by a high-intensity pulsed electron beam of submillisecond exposure time, electron beam energy 25 J/cm² (arrows indicate the direction of exposure to the electron beam).

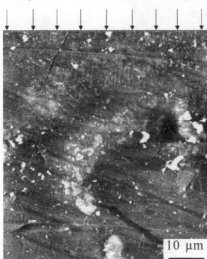

1 – directionally recrystallized grains located in the treated layer, 2 – not melted grain boundary, 3 – grain body.

Fig. 4.20. Atomic force microscopy of a sample treated with a high-intensity pulsed electron beam, submillisecond exposure time, electron beam energy 30 J/cm² (arrows indicate the direction of exposure to the electron beam)

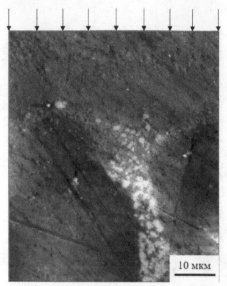

*1 – directionally recrystallized grains located in
the treated layer, 2 – intergranular eutectic, 3,
4 – dendritic grains*

Fig. 4.21. Al–Si atomic force microscopy provides a submillisecond exposure time processed by a high-intensity pulsed electron beam and an electron beam energy of 35 J/cm² (arrows indicate the direction of exposure to the electron beam).

The results obtained using atomic force microscopy make it possible to establish that the regimes with the energetic action of an electron beam are from 25 to 35 J/cm². The presented modes are characterized by a fine-grained cellular structure, and also have the smallest roughness of the treated layer (17–33 nm) and the substrate (45–57 nm), in comparison with other modes. The choice of optimal processing modes according to the results of atomic force microscopy correlates with the selected modes according to the results of microhardness measurements.

4.2.2 Comparison of the results of atomic force microscopy of samples treated with electron beams and a multiphase plasma jet of the Al–Y₂O₃ system

A comparison was made of the results of atomic force microscopy studies treated with PMPJ with samples treated with IPEB. Upon detection of defects – micropores were found in the coating in the first method surface treatment of the sample. When processed by an electron beam, a fine-grained cellular structure is formed in the

Table 4.4. Roughness of the coating and substrate of samples subjected to IPEB under various conditions

Density of energy of electron beam E_s, J/cm²	Average roughness of treated layer Ra, nm	Average roughness of substrate Ra, nm
10	99	77
15	30	30
20	27	57
25	27	45
30	33	51
35	17	45

treated layer; there are no defects in the form of micropores. The roughness Ra of the coating of samples treated with PMPJ varies from 50 to 270 nm, the roughness of the substrate near the coating is 50–90 nm. The roughness of the treated layer of samples after the IPEB is in the range from 17 to 90 nm, the substrate near the treated layer is from 30 to 77 nm. Thus, the average roughness of the samples treated with IPEB is less than the samples subjected to PMPJ. IPEB modifies surface layers. Al–Si provides the creation of a fine-grained and cellular surface, as well as the creation of a surface with a high degree of roughness.

The results presented in the section were published in [14, 15].

4.3. Analysis of changes in the fine structure and phase composition of the surface layers of Al–Si alloys subjected to irradiation with an intense pulsed electron beam

4.3.1. Analysis of the Al–Si structure in cast state

Typical images of the structure of the etched thin section obtained by scanning electron microscopy are shown in Fig. 4.22.

It is clearly seen that, especially at relatively large magnifications (Fig. 4.22 b), which are studied in the work, it is a multiphase aggregate, the structure of which includes grain products based on aluminium, grain Al–Si eutectics, which include primary nitrogen atoms and intermetallic compounds, sizes and shapes that vary over a wide range. The presence of intermetallic compounds leads to a decrease in crack resistance [16–19]. The presence of micropores (Fig. 4.22 c, micropores are indicated by arrows).

Using X-ray microspectral analysis methods (mapping method), it was found that the main alloying elements are nickel, magnesium and copper (Fig. 4.23).

X-ray spectral analysis allows studies of elemental composition. Figure 4.24 and Table 4.5 present the results of such an analysis. It can be seen from the table that the alloying elements are distributed very nonuniformly over the material, forming compounds that differ in the elemental composition.

The elemental composition of the inclusions shown in Fig. 4.24 *a*, indicates a very heterogeneous distribution of alloying elements in the cast alloy.

The bulk of the material (the area indicated by the number 1) is represented by a solid solution based on aluminium, the content of alloying elements in which is below the average of that in Fig. 4.22 *a*. The inclusions indicated in Fig. 4.24 *a*, numbers 2, 3 and 4, are enriched with magnesium, silicon and copper (inclusion 2), copper and magnesium (inclusion 3), nickel, iron and copper (inclusion 4). Similar results indicating the inhomogeneous distribution of alloying elements in the molten state were obtained in the analysis of many

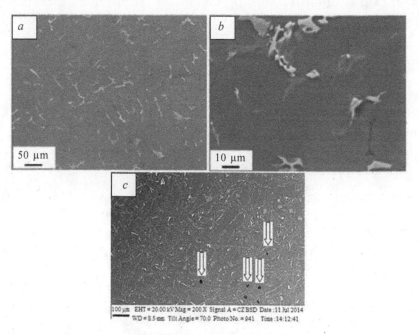

Fig. 4.22. The structure of the location in the molten state (on (*c*) arrows indicate micropores). SEM.

other inclusions. The presence of micropores (Fig. 4.22 *c*, micropores are indicated by arrows).

The phase composition of the material also determines the methods of x-ray diffraction analysis. The plot of the x-ray obtained from the test material is shown in Fig. 4.25.

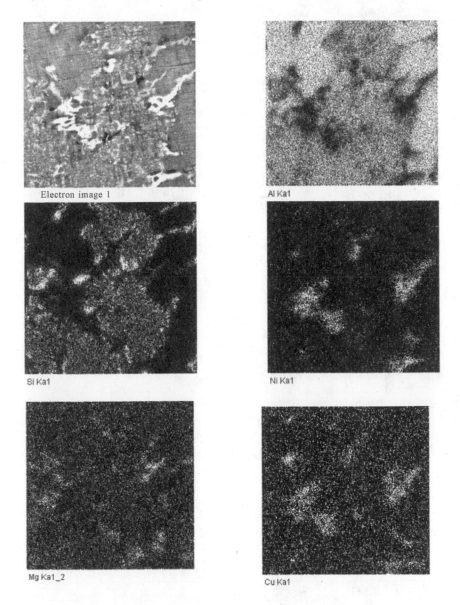

Fig. 4.23. Distribution maps of alloying elements over the area of the initial sample, electron microscopic image that is used on (*a*).

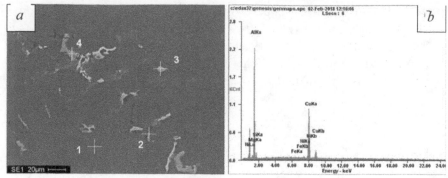

Fig. 4.24. Electron microscopic image of the structure of the cast state (*a*) and energy spectra (*b*), the surface dimensions of the plot shown in (*a*). SEM.

The results of the quantitative X-ray analysis are given in Table 4.6. The results of the study should be presented in the form of solid solutions based on aluminium and nitrogen. The crystal lattice parameters of aluminium and nitrogen in the alloy under study are close to the crystal lattice parameters of pure elements, which indicate the separation of these elements during crystallization. X-ray phase analysis also revealed the aluminium element of the composition AlCu$_3$ ($\approx$3.5 rel.%), the diffraction maxima of which are shown in Fig. 4.25 (arrows).

4.3.2. Evolution of the structures of Al–Si irradiated by an intense electrom beam of different density and examined by scanning electron microscopy

Homogenization of the surface layer of the material is carried out by irradiating the surface with an intense pulsed electron beam (IPEB). Irradiation with an electron beam with an energy density $E_s = 10$ J/cm^2 is accompanied by smooth motion in all regions with inclusions of the second stage (Fig. 4.26). The width of the molten

Table 4.5. The results of X-ray microspectral analysis of a section of a thin section, electron-microscopic image shown in Fig. 4.24. The results are presented in wt. %

Region	Mg	Al	Si	Fe	Ni	Cu
1	0,97	94.97	1.60	0.33	0.43	1,70
2	22.71	47.87	25.79	0.25	0.25	3.13
3	1.67	62.87	3.76	0.50	0.56	30.65
4	0.51	68.81	0.73	4.39	22.56	3.00

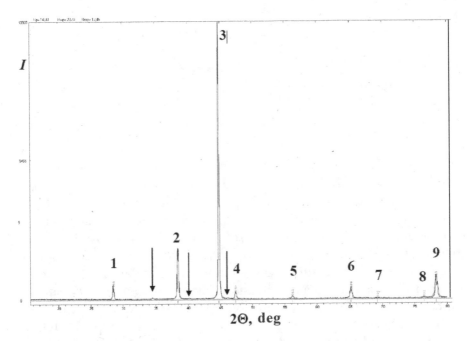

Fig. 4.25. X-ray plot of the studied material; the numbers indicate the diffraction maxima of aluminium and nitrogen: 1 – (111) Si; 2 – (111) Al; 3 – (200) Al; 4 - (220) Si; 5 – (311) Si; 6 – (220) Al; 7 – (400) Si; 8 – (331) Si; 9 – (311) Al. The arrows indicate the diffraction lines of the AlCu$_3$ phase.

layer (along the irradiated surface) reaches (10–15) μm (Fig. 4.26 *b*). The reason for the selective melting of the surface is a low coefficient of thermal conductivity, which leads to overheating with a doped layer of aluminium during pulsed irradiation of samples. In areas of the specimen free from inclusions there is no melting of the material, as indicated by the scratches remaining after mechanical polishing of the material (Fig. 4.26 *b*).

An increase in the energy density of the electron beam to 15 J/cm^2 leads to the disappearance of a scratch on the surface of the sample, which indicates melting of the surface layer of aluminium (Fig. 4.27 *a*). The melting of the volume of aluminium accompanying the formation of the second stage is accompanied by the formation of the structure of cellular crystallization; the cell size is 200–220 nm (Fig. 4.27 *b*). The width of the interlayer with the structure of cellular crystallization provides (15–20) μm. In the second case, it cannot be revealed in the cell that this may indicate a small thickness of the molten surface layer.

Table 4.6. The results of X-ray diffraction analysis of the sample in the cast state The results are presented in wt. % (a_0 - tabulated value, a – value in the alloy)

Phase	Content, rel.%	Lattice	Lattice parameter, nm		Atomic radius, nm
			a_0	a	
Al	84.2	$Fm3m$	0.4050	0.40484	0.143
Si	12.3	$Fm3ms$	0.54307	0.54265	0.132

At an electron beam energy density of 20 J/cm² or more, the observed active dissolution includes an intermetallic phase that develops in the surface layer (Fig. 4.28). Due to the high cooling rate of the surface layer, elements are formed that form intermetallic compounds, which form a solid solution in the bulk of aluminium. These areas have a lighter contrast (Fig. 4.28 *b*). The dissolution of the included intermetallic compounds under ultrafast cooling of the molten layer is accompanied by the formation of networks of surface microcracks, which may indicate a high level of tensile stress and increased brittleness of these material particles (Fig. 4.28 *c*). Melting and high-speed cooling of the surface layer leads to the formation of a cellular crystallization structure (Fig. 4.28 *d*). The cell sizes vary within 400–500 nm and slightly increase with increasing electron beam energy density in the range (20–35) J/cm².

Figure 4.29 shows the structure of transverse sections. It can be seen that the thickness of the surface layer without intermetallic particles varies from 35 μm to 80 μm and increases with increasing electron beam energy density from 25 J/cm² to 35 J/cm² (Fig. 4.29 *a*). In the volume of this surface layer there are uniformly distributed inclusions, the sizes of which vary within (150–175) nm

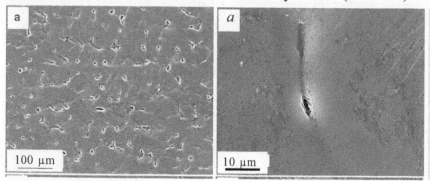

Fig. 4.26. An electron microscopic image of the surface structure of a sample subjected to electron beam irradiation at an electron beam energy density of 10 J/cm².

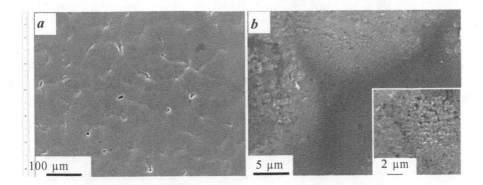

Fig. 4.27. Electron microscopic image of the surface structure of a sample irradiated with an electron beam at an electron beam energy density of 15 J/cm^2.

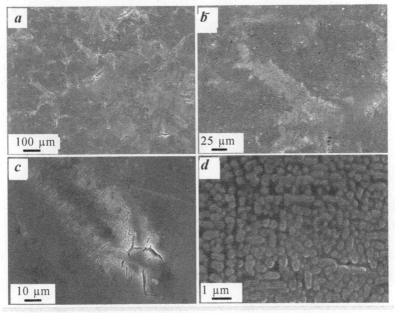

Fig. 4.28. Electron microscopic image of the surface structure of the sample irradiated with an electron beam at an electron beam energy density of 20 J/cm^2.

(Fig. 4.29 *b*). Judging by the lighter (relative to the matrix) contrast of these inclusions, we can conclude that they are enriched with atoms of alloying and impurity elements having a greater atomic weight compared to aluminium and, therefore, are particles of the intermetallic phase.

According to the results presented in this section, we can conclude that the optimal parameters of electron-beam processing, as in the

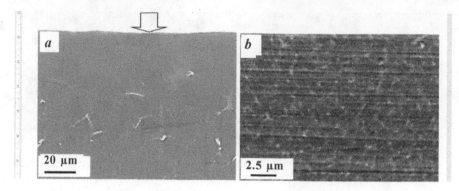

Fig. 4.29. Electron microscopic image of the cross-sectional structure of a sample irradiated with an intense pulsed electron beam at an electron beam energy density of 25 J/cm² (*a*) and 35 J/cm² (*b*). The arrow on (*a*) indicates the irradiated surface.

case of microhardness and tribological characteristics, are regimes with an electron beam energy density of 20 J/cm² to 35 J/cm². In this regard, further, detailed studies of the structure and phase composition were performed for regimes with an electron beam energy density of 25 and 35 J/cm².

4.3.3. Modification of the structure of an Al–Si alloy by an intense pulsed electron beam with an energy density of 25 J/cm²

Irradiation of the surface of the samples with an intense pulsed electron beam in the melting mode (25 J/cm²; 150 μs; 3 pulses) is accompanied by the formation of a cellular crystallization structure (Fig. 4.30). Inclusions of silicon and intermetallic compounds after irradiation with an electron beam are not observed, which, obviously, indicates their melting.

Studies of the elemental composition of the surface layer after electron beam irradiation performed by X-ray spectral analysis revealed a significant decrease in the concentration of silicon in the surface layer of the irradiated samples (Fig. 4.31). By analyzing the results presented in Fig. 4.31, it can be stated that in the surface layer of the sample after irradiation, a significant, 1.5–2 times decrease in the concentration of alloying elements is observed (Fig. 4.31 *b*).

The phase composition and state of the crystal lattice of the surface layer of the material irradiated with an electron beam was studied by X-ray diffraction analysis. The research results presented in Table 4.7 indicate the presence of three phases in the surface

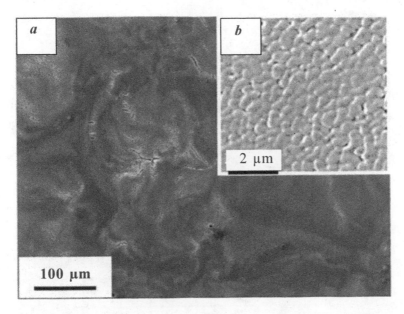

Fig. 4.30. Structure of an Al–Si alloy after irradiation with an intense pulsed electron beam. SEM.

Table 4.7. Results of X-ray diffraction analysis of a sample irradiated by an intense pulsed electron beam

Phase composition	Relative content, %	Lattice parameter, nm	
		irradiated	tabular [6]
AlSi	53.13	0.40412	
Al	38.29	0.40419	0.40494
Si	8.58	0.54191	0.54307

layer. The crystal lattice parameters of the detected phases differ from the tabulated values for pure elements. This fact may indicate the formation of solid solutions based on aluminium and silicon, which occurs during high-speed crystallization of the molten layer.

Smaller values (relative to tabular values) of the crystal lattice parameter of aluminium, silicon, and the AlSi phase can indicate doping of these phases with elements whose atom sizes are smaller than the sizes of aluminium or silicon atoms: $R(\text{Al}) = 0.143$ nm; $R(\text{Si}) = 0.132$ nm; $R(\text{Cu}) = 0.128$ nm; $R(\text{Fe}) = 0.126$ nm; $R(\text{Ni}) = 0.124$ nm [19].

The electron microscopic image of the cross-sectional structure are shown in Fig. 4.32 and it can be said that radiation occurs with

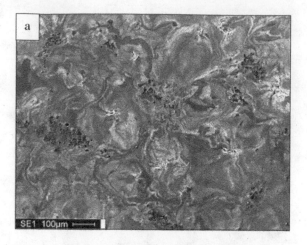

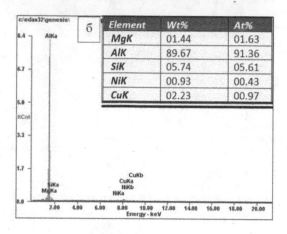

Element	Wt%	At%
MgK	01.44	01.63
AlK	89.67	91.36
SiK	05.74	05.61
NiK	00.93	00.43
CuK	02.23	00.97

Fig. 4.31. Electron microscopic image of the structure of the material after irradiation (*a*); (*b*) shows the energy spectra obtained from this section.

the help of an intense electron pulse. 25 J/cm^2; 150 μs; 3 pulses), vary within 40–60 μm.

The methods of transmission electron microscopy of thin foils made from the cross section of samples were used.

Figure 4.33 presents characteristic images of a multiphase submicron structure in which structures are formed that are crystalline structures formed as a result of melting (25 J/cm^2, 150 μs, 3 pulses), accompanied by high-speed crystallization.

As a rule, two types of cells are formed. First of all, cells with no second phase precipitates in the volume. In some cases, cases, nanoscale particles of a rounded shape, located randomly, are observed. Secondly, cells in which lamellar eutectic structures are present (Fig. 4.33 *a*). Note that the cells of the first type in this mode are predominantly typical structures of the surface layer with a thickness of ≈10 μm. At a greater distance from the surface, a mixed type structure is formed, represented by cells of the first and second type. The sizes of the interlayers do not exceed 100 nm (Fig. 4.34 *b*).

The methods of X-ray microanalysis of thin foils were used to study the elemental composition of the cells of the first and second type.

It was found that the volume of cells of the first type is enriched in aluminium atoms (Fig. 4.34 *a*), i.e. is an aluminium based solid solution. The cells of the second type were formed in the form of two different groups: enriched with aluminium atoms (Fig. 4.34 *a*) and nitrogen atoms (Fig. 4.34 *b*), i.e. represent eutectic cells equipped

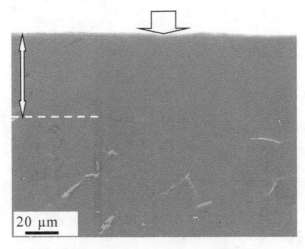

Fig. 4.32. An electron microscopic image of the structure of a transverse section irradiated with an intense pulsed electron beam. The arrows indicate the irradiated surface and the thickness of the surface layer in which primary inclusions of the second phase are not detected by SEM methods.

with Al–Si. The transverse dimensions of the layers of nitrogen and aluminium vary in the range of 40–60 nm. Interlayers of the second level, located at the interface, are enriched with the atoms of nitrogen, copper, nickel and iron (Fig. 4.35).

In quantitative terms, the results of elemental analysis of this foil section are given in Table 4.8. It is clearly seen that silicon and copper are the main elements of the analysis of the section are aluminium, silicon and copper. A small amount of magnesium, nickel,

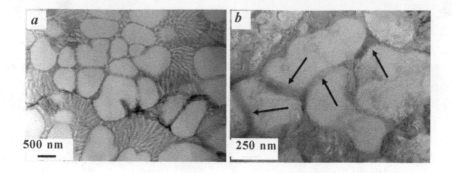

Fig. 4.33. Electron microscopic image of the cellular crystallization structure of a sample of the surface layer irradiated by an intense pulsed electron beam (25 J/cm², 150 µs, 3 pulses). Interlayers are indicated on the second (second) arrow.

iron and titanium and manganese was also found. Similar results were obtained using SEM and X-ray phase analysis.

The average size of the crystalline layer in size indicates that the average cell size is 0.4 µm ± 0.11 µm. The size distribution of cells is monomodal, which indicates a high dimensional uniformity of the generating structure (Fig. 4.36).

At a greater distance from the surface, the apparent average crystal sizes increase, and on the surface of the lower layer with the structure of the cellular structure, values of 0.65 µm ± 0.22 µm are reached.

The TEM methods used included the analysis of microdiffraction patterns, bright-field and dark-field images.

Figure 4.37 shows the results of electron microscopic dark-field studies of the structure of lamellar eutectics. The analysis of the microelectron diffraction pattern (Fig. 4.37 c) gives reason to conclude that the plates (Fig. 4.37 d) are formed by silicon. Si plates are polycrystals ranging in size from 5 to 10 nm. The nanocrystalline structure of the Si plate is also indicates bythe ring structure of the microelectron diffraction patterns (Fig. 4.37 c).

Figure 4.38 presents the results of dark-field electron microscopy analysis of the structure of cellular crystallization. Analysis of the microelectron diffraction pattern shown in Fig. 4.38 c indicates that high-speed crystallization cells are formed by an aluminium-based solid solution (Fig. 4.38 b). Interlayers separating crystallization cells are multiphase formations. The analysis of microelectron diffraction patterns obtained from the volumes of the foil containing

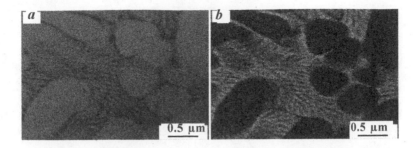

Fig. 4.34. The structure of the surface layer of the material obtained in X-ray radiation of aluminium (*a*) and nitrogen atom (*b*).

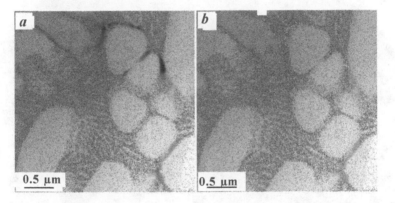

Fig. 4.35. The structure of the foil portion formed as a result of the light signals (*a*); *b* – bright field images and images obtained in X-rays.

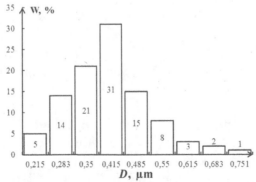

Fig. 4.36. Cell distribution in high-speed crystallization of the surface layer of a material irradiated by an intense pulsed electron beam.

the interlayers revealed particles of the following phases in the interlayers: $Cu_{15}Si_4$ (Fig. 4.38 *d*), silicon (Fig. 4.38 *e*), as well as copper and Al_4Cu_9.

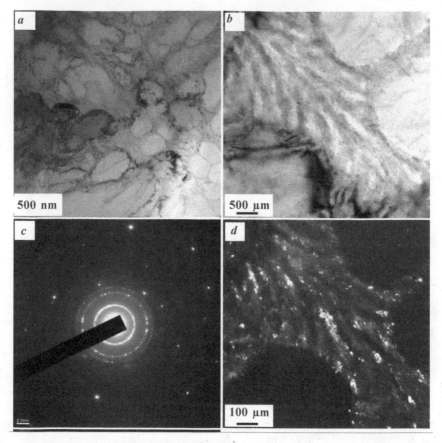

Fig. 4.37. Electron microscopic image of the structure of the alloy formed as a result of irradiation with an intense pulsed electron beam; *a, b* – bright-field images; *c* – microelectron diffraction pattern; *d* is the dark field obtained in the first diffraction ring of [111] Si.

The layer-by-layer analysis of the structures formed by TEM methods showed that the surface layer formed upon irradiation has a gradient structure (Fig. 4.39).

A layer with a thickness of 8–10 μm adjacent to the irradiated surface has a cellular structure, the cell boundaries are separated by interlayers of the second phase, the thickness of which does not exceed 100 nm (Fig. 4.39 *a*). There are no eutectic grains.

At a greater distance from the irradiated surface, cells (grains) with a lamellar substructure (eutectic) are detected in the structure of cellular crystallization (Fig. 4.39 *b–d*). The relative content of such grains increases with increasing distance from the irradiated surface. The first eutectic grains are found in a layer located at a depth of ≈15 μm. As the distance from the irradiated surface, the relative

Table 4.8. The results of xX-ray spectral analysis

Mass fraction of elements, %							
Al	Si	Cu	Ni	Mg	Ti	Fe	Mn
86.60	6.16	4.92	0.76	1.11	0.13	0.29	0.04

content of eutectic grains increases. Eutectic grains are located in islands or interlayers between cells of high-speed crystallization of aluminium. The presence of eutectic grains indicates the existence of local regions in the surface layer of the material with a relatively high (≈12 at.%) concentration of silicon atoms. The sizes of eutectic grains are close to the sizes of grains of a solid solution based on aluminium (crystallization cells). The transverse dimensions of the eutectic plates vary from 25 nm to 50 nm.

Inclusions of cast intermetallic compounds located in the structure of cellular crystallization are found at a depth of (50–70) μm. Inclusions of intermetallic compounds act as centres of cellular crystallization.

A layer of material in which only aluminium melts and primary inclusions of silicon and intermetallic compounds are present is detected at a distance of 80–90 μm from the irradiated surface. In this case, cells with high-speed crystallization of aluminium are observed in the structure. There are no grains of lamellar eutectic of submicron sizes. At a distance of (100–120) μm from the irradiated surface, the structure of cellular crystallization is not detected (Fig. 4.39 e, f).

The mapping method revealed the distribution of aluminium and alloying elements in the structure of the irradiated layer of the sample. It was found that the cell volume is enriched with aluminium atoms, i.e. formed by an aluminium-based solid solution (Fig. 4.40 a, d). The layers separating the cells are enriched mainly by silicon atoms (Fig. 4.40 b, e) and copper (Fig. 4.40 c, f).

The elemental composition of the material irradiated by the electron beam naturally varies depending on the distance from the irradiated surface (Fig. 4.41). Analyzing the results shown in Fig.4.41, it can be noted that the silicon concentration changes most significantly, the relative content of which increases from 6.2 wt. % in the surface layer to 10.4 wt. % in a layer located at a depth of 30 μm. Thus, melting of the surface layers of the material by an intense pulsed electron beam is accompanied by a decrease in the concentration of silicon in the surface layer with a thickness of up to 30 μm.

4.3.4. Analysis of structural changes in the Al–Si alloy irradiated by a pulsed electron beam with an energy density of 35 J/cm²

The scanning electron microscopy study of the surface of samples irradiated with an intense pulsed electron beam with parameters of 35 J/cm², 150 μs, 3 pulses. It did not show significant differences from the structure formed upon irradiation with parameters of 25 J/cm², 150 μs, 3 pulses. However, there are some features that need to be considered. It was established that the dimensions of the melted layer increase with increasing energy density of the electron beam. In a layer with a thickness of 50–100 μm, a cellular crystallization structure is formed that is characteristic of rapidly quenched material (Fig. 4.42), the crystallite size of which varies from 350 nm to 550 nm.

At an electron beam energy of 35 J/cm², the thickness of the surface modified layer in which the primary inclusions of silicon and intermetallic compounds cannot be detected by SEM methods increases and becomes equal to from 70 μm to 100 μm (Fig. 4.43 a).

An analysis of transverse sections irradiated by an electron beam revealed the formation of a multilayer gradient structure. According to the morphology of the defective substructure, three layers can be conditionally distinguished, which in this paper are called surface (Fig. 4.43 a, layer 1), transition (Fig. 4.43 a, layer 2) and a heat-affected layer (Fig. 4.43 a, layer 3). The surface layer has a honeycomb crystallization structure formed during high-speed cooling of the material from the molten state (Fig. 4.43 b, layer 1). In this layer the SEM methods cannot identify the primary inclusions of the second phase. The transition layer is characterized by the presence of primary inclusions of the second phase, which are the centers of crystallization of aluminium (Fig. 4.43 b, layer 2).

Figure 4.44 and Table 4.9 shows the results of X-ray spectral analysis of the surface and transition layers of the sample. Analyzing the presented results, one can note a relatively low concentration of silicon and an increased concentration of nickel and copper, relative to the initial state. Therefore, high-speed crystallization is accompanied by a redistribution of alloying elements at the crystallization front, namely, the displacement of silicon from the crystallizing layer. The transition layer, as noted above, is characterized by the presence of incompletely dissolved inclusions of the intermetallic phase (Fig. 4.44 *a*, region 3) and is substantially enriched with silicon (Fig. 4.44 *a*, region 2).

The phase composition and state of the crystal lattice of aluminium, modified by an intense pulsed electron beam, was studied by X-ray diffraction analysis. Figure 4.45 presents a plot of the x-ray obtained from the irradiated sample; the results of a quantitative analysis of the x-ray are given in Table 4.10.

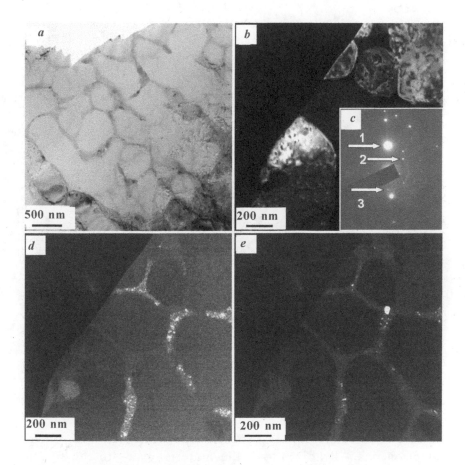

Fig. 4.38. An electron microscopic image of the structure of the alloy formed as a result of irradiation with an intense pulsed electron beam; *a* – bright field image; *c* – microelectron diffraction pattern; *b, d, e* – dark fields obtained in reflections of [111] Al, [321] $Cu_{15}Si_4$, [220] Si, respectively. On (*c*) the arrows indicate the reflexes in which the dark fields are obtained: 1 – (*b*), 2 – (*d*), 3 – (*e*).

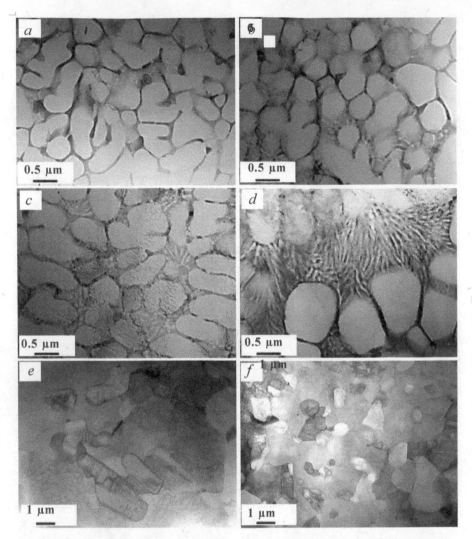

Fig. 4.39. Electron microscopic image of an Al–Si alloy processed by an intense pulsed electron beam; *a* – structure of a layer 5 μm thick adjacent to the irradiated surface; *b* – the structure of the layer located at a distance of X = 15 μm from the irradiated surface; *c* – the structure of the layer at X = 30 μm; *d* – at X = 50 μm; *e* – at X = 120 μm; *f* – at X = 200 μm; TEM.

Analyzing the results given in Table 4.10, it can be stated that electron beam irradiation of the material leads to the formation of two aluminium-based solid solutions, indicated in the table as AlSi and Al, and the precipitation of copper aluminide of the composition $AlCu_2$ and silicon. The crystal lattice parameter of the AlSi solid solution is less than the crystal lattice parameter of pure aluminium, equal to 0.40494 nm [19]. This is due to the fact that

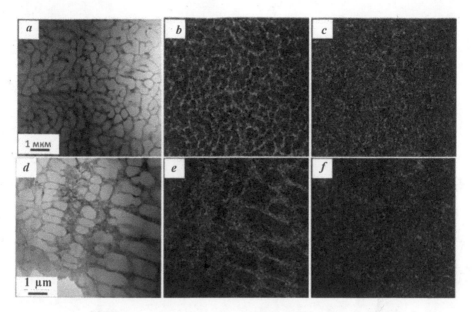

Fig. 4.40. Electron microscopic image of the structure of the alloy irradiated with an electron beam (*a, d*); on (*b, c, d, e*) are images of the alloy structure obtained in x-ray radiation of silicon atoms (*b, e*) and copper atoms (*c, f*); *a–c* – a layer located at a depth of ≈8 μm; (*d*) – at a depth of ≈30 μm.

the atomic radius of silicon (0.132 nm) is smaller than the atomic radius of aluminium (0.143 nm) [19] and, therefore, the substitution of aluminium atoms by silicon will lead to a decrease in the crystal lattice parameter of the AlSi solid solution. The crystal lattice parameter of the second phase based on an aluminium solid solution is greater than the crystal lattice parameter of pure aluminium. This is due to the dissolution of intermetallic particles and the enrichment of the solid solution with metal atoms whose atomic radius is greater than that of aluminium. The crystal lattice parameter of silicon precipitates is less than the tabulated value of the crystal lattice parameter (*a* (Si) = 0.54307 nm) [19]. This means that during the crystallization process a silicon-based solid solution is formed, in which copper, nickel and iron atoms can be present, because the atomic radii of these elements are smaller than the atomic radius of silicon. The relative silicon content in the modified layer is relatively small and close in magnitude to the value obtained by X-ray microanalysis.

Layer-by-layer analysis of the defective substructure of the modified layer was performed by TEM methods (Fig. 4.46). Note

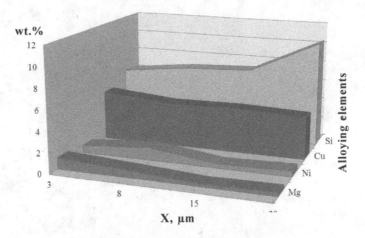

Fig. 4.41. Dependence of the concentration of alloying elements on the distance to the irradiated surface.

that the surface layer with a thickness of up to 100 μm has a cellular structure of high-speed crystallization. The cell size varies from 400 nm to 600 nm. The cells are separated by interlayers of the second phase, the thickness of which varies from 80 nm to 200 nm. As you move away from the irradiated surface, the cell sizes increase, reaching 1.0–1.2 μm (Fig. 4.46 *b*).

Transmission electron microscopy of transverse foils

An increase in the size of crystallization cells is accompanied by the transformation of interlayers located along their boundaries into spherical particles (Fig. 4.46 *c*). It was found that the average cell size increases in the range from 0.3 μm to 0.6 μm with an increase in the energy density of the electron beam (25–35 J/cm²) with the other radiation parameters unchanged. At a distance of 120–130 μm from the irradiated surface, primary particles of silicon and intermetallic compounds are detected (Fig. 4.46 *d*).

Table 4.11 shows the elemental composition of various sections of the surface layer of an alloy irradiated with an electron beam.

Analyzing the results presented in Table 4.11, it can be noted that irradiating the material with an intense pulsed electron beam in the melting mode of the surface layer leads, firstly, to homogenization of the elemental composition of the surface layer; secondly, to a decrease in the concentration of silicon atoms in the surface layer, which increases with increasing energy density of the electron beam.

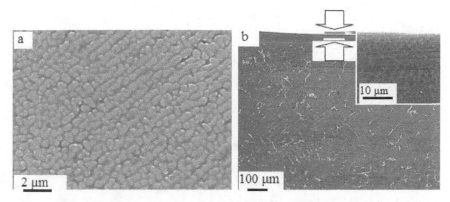

Fig. 4.42. The structure of the alloy irradiated by an intense pulsed electron beam (35 J/cm², 150 μs, 3 pulses); *a* – surface structure of the irradiation; *b* – the structure of the transverse section (lines and arrows indicate the layer of high-speed crystallization). The tab shows the structure of cellular crystallization of the cross section of the surface layer.

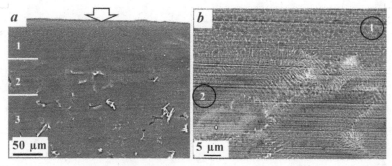

Fig. 4.43. An electron microscopic image of the structure of a transverse section of a sample irradiated by an intense pulsed electron beam. The arrow on (*a*) indicates the irradiated surface; the numbers indicate the layers: 1 – surface; 2 – transitional; 3 – heat-affected layer.

Irradiation of an Al–Si alloy by an intense pulsed electron beam is accompanied by the formation of a gradient structure – the structure of cellular crystallization passes, with distance from the processing surface, into a mixed type structure, in which partially dissolved inclusions of foundry origin are present along with the cells (Fig. 4.47). Analyzing the structure of this transition layer, we can note the absence of plate-like inclusions in it. In most cases, inclusions have a quasi-equiaxial shape (Fig. 4.47 *a*). This statement is true both for silicon particles and for particles of intermetallic compounds. It should be noted that globularization of silicon particles and intermetallic compounds should significantly increase the plastic

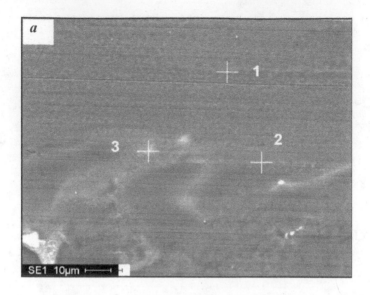

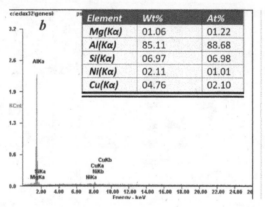

Element	Wt%	At%
Mg(Kα)	01.06	01.22
Al(Kα)	85.11	88.68
Si(Kα)	06.97	06.98
Ni(Kα)	02.11	01.01
Cu(Kα)	04.76	02.10

Fig. 4.44. Electron microscopic image of the structure of the initial state irradiated by an electron beam (*a*) and energy spectra (*b*) obtained from the surface area shown in (*a*). In (*b*) the table shows the quantitative results of the analysis of the elemental composition of the site, indicated in (*a*) by "1"

properties of not only the modified layer, but also of the material as a whole.

SEM and TEM methods have established that irradiation of an Al–Si alloy with an intense pulsed electron beam, as a result of high-speed melting and crystallization, leads to the formation of a multilayer gradient structure, the surface layer of which is up to 100 μm thick, is free from primary inclusions of silicon and intermetallic compounds and has a cellular crystallization structure. Crystallization

Table 4.9. The results of the micro X-ray spectral analysis of the section of the section, the electron-microscopic image of which is shown in Fig. 4.23

Element	Number of analyzed section					
	1		2		3	
	wt%	at%	wt%	at%	wt%	at%
Mg	01.06	01.22	01.24	01.43	02.66	03.26
Al	85.11	88.68	76.42	79.09	72.51	80.00
Si	06.97	06.98	17.37	17.40	08.34	08.84
Ni	00.88	00.42	04.44	02.25	04.44	02.25
Cu	04.75	02.10	04.09	01.66	12.05	05.65

Table 4.10. Results of X-ray diffraction analysis of a sample irradiated by an intense pulsed electron beam (CSR – coherent scattering region)

Phase composition	Relative content, %	Lattice parameter, nm		D (CSR), nm	$\Delta d/d$, 10^{-3}
		a	c		
AlSi	40.8	0.40435			
Si	4.1	0.54274			
Al	40.7	0.40508		34.48	2.311
AlCu$_2$	14.4	0.40311	0.57492	29.83	0.926

cells (solid solution based on aluminium) are separated by nanosized interlayers of the second phase enriched with alloying elements (silicon, copper, nickel). X-ray diffraction analysis revealed the formation, as a result of high-speed crystallization, of a modified alloy layer of aluminium-based alloys enriched with silicon atoms and other alloying elements, a silicon-based solid solution, particles of copper aluminide of the composition AlCu$_2$. It is shown that a significant transformation of the surface layer that occurs when the material is irradiated with an intense pulsed electron beam is accompanied by a multiple increase in the wear resistance and microhardness of the modified layer.

4.3.5. Comparison of changes in the phase composition and structure of an Al–Si alloy subjected to electron-beam treatment and multiphase plasma jet by the Al–Y$_2$O$_3$ system

Summarizing the studies of the morphology of the structure of the Al–Si alloy subjected to surface treatment by the methods of PMPJ and IPEB, we can distinguish characteristic features. The coating of the Al–Y$_2$O$_3$ system, formed on the surface of the samples by

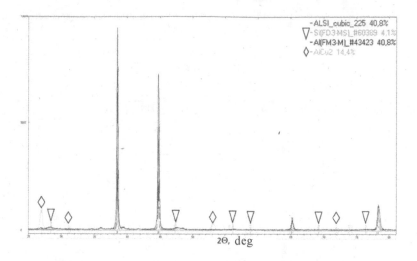

Fig. 4.45. X-ray plot of a sample irradiated by an intense pulsed electron beam/

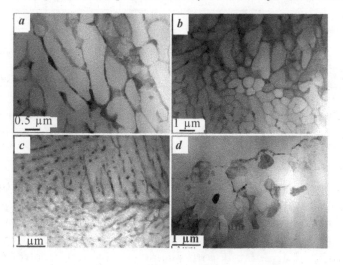

Fig. 4.46. An electron microscopic image of the structure of an alloy irradiated with an intense pulsed electron beam; *a* – structure of a layer adjacent to the irradiated surface, *b* – a layer located at a distance of 70 μm from the irradiated surface, *c* – 100 μm from the irradiated surface, *d* – 120 μm from the irradiated surface.

the PMPJ method, is generally characterized by high porosity with an inhomogeneous distribution over the thickness, which depends on the treatment mode and varies from 30 to 130 μm, and an inhomogeneous content of alloying elements. It should be noted that the main elements of the modified surface are aluminium, yttrium and titanium. As a result of processing the alloy surface with intense electron beams, a homogenized layer is formed, the thickness of which increases with increasing electron beam density and is: 24 μm for an energy density of 10 J/cm², 38 μm for 15 J/cm², 42 μm for 20 J/cm², 61 μm for 25 J/cm², 66 μm for 30 J/cm² and 100 μm for 35 J/cm².

Studies were carried out on the defective substructure of the material. It was found that surface modification of the cast Al–Si alloy leads to the formation of cellular crystallization structures in the surface layer: the size of crystallization cells varies between 200–450 nm for processing by PMPJ, and within 200–500 nm for processing by electron beams.

In PMPJ, the hardening of the surface layers of a material is achieved due to the formation of a multilayer, multiphase nanocrystalline structure formed mainly by oxides and silicates of aluminium and yttrium. In the case of IPEB, hardening occurs as a result of high-speed melting and crystallization, which also leads to the formation of a multilayer gradient structure, the surface layer of which is up to 100 μm thick, depending on the processing parameters, is free from primary inclusions of silicon and intermetallic compounds, and due to the formation of solid solutions on based on aluminium, enriched with silicon atoms and

Table 4.11. The elemental composition of various regions of the cellular substructure of the surface layer of a sample irradiated by an intense electron beam

Element	Concentration, %								
	Number of section of X-ray spectrum microanalysis								
	*1/25	2/25	3/25	4/25	5/25	6/25	7/25	8/25	9/25
Mg	0.32	0.81	0.44	1.28	0.0	0.14	0.0	0.0	0.0
Al	90.14	89.41	86.83	89.82	92.14	90.97	91.77	91.1	92.83
Si	7.15	6.88	10.36	6.13	5.86	6.25	6.17	3.68	4.69
Ti	0.11	0.13	0.11	0.08	0.04	0.05	0.01	0.13	0.25
Mn	0.02	0.01	0.02	0.02	0.0	0.01	0.0	0.04	0.02
Fe	0.11	0.24	0.12	0.14	0.14	0.07	0.12	0.59	0.1
Ni	0.2	0.6	0.28	0.36	0.37	0.61	0.41	2.02	0.45
Cu	1.94	1.9	1.83	2.17	1.45	1.9	1.53	2.44	1.67

* The number of the analyzed section and the energy density of the electron beam are indicated

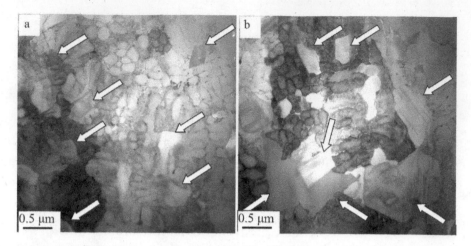

Fig. 4.47. The structure of the alloy irradiated by an intense pulsed electron beam (35 J/cm², 150 µs, 3 pulses); *a* – layer at a depth of 70 µm; *b* – layer at a depth of 90 µm. Arrows indicate inclusions in the cast state/

other alloying elements, a solid solution based on silicon, particles of copper aluminide of the composition $AlCu_2$.

The coating formed by the PMPJ method was characterized by the presence of a large number of defects. Among which one can distinguish porosity and a large number of microcracks, and heterogeneity in the thickness and distribution of alloying elements. The alloy surface treated by electron beams was characterized by the formation of a homogeneous homogenized layer, the thickness of which varied from the processing parameters. In the course of the work, optimal spraying modes of coatings and optimal processing regimes by intense electron beams were established. Also from the literature review and from the analysis of experimental data, it was found that using IPEB it is possible to reduce the surface defects of the material and coatings applied to the substrate. Based on this, further research goals and objectives have been identified, namely, the combined treatment of the Al–Si alloy with PMPJ methods with further IPEB of the resulting coatings.

4.4. Theoretical studies of the effects of low-energy high-current electron beams on Al–Si alloys

One of the most promising methods for modifying the structure of the surface layer of products from various materials in order to increase their operational characteristics is currently the IPEB. This direction is developing on the basis of the constructed installations SOLO [20–22], GESA [23–25] and NADEZHDA [26–28]. The effect of electron beams on materials is currently a complex set of phenomena, which includes heating, melting, convective flows in the liquid layer, evaporation of matter and subsequent crystallization [29–33]. The structure, phase composition, defective substructure and mechanical properties of the processed materials will depend on how these processes proceed. Electron microscopic studies [34–36] of the surface layers of materials show that the action of electron beams with an absorbed power density of 10^4–10^5 W/cm^2 leads to the formation of cellular crystallization structures ranging in size from ~100 nm to 1 μm. In this case, the mechanism of the appearance of a cellular structure on the melt surface can be associated with the development of thermocapillary instability [37], which is formed due to the action of thermocapillary forces along the melt surface due to the dependence of surface tension on temperature and the presence of a stationary temperature gradient. In works on the theoretical study of the thermocapillary instability of molten layers of materials under the action of laser radiation [38–47], a viscous liquid is considered that occupies a stationary state layer on the free surface of which heat q is absorbed. The linearization of the Navier–Stokes equations and the energy equation with the corresponding boundary conditions leads to a boundary-value problem for perturbations with zero boundary conditions. The requirement of a nonzero solution to this boundary value problem leads to an algebraic equation – a dispersion equation that relates the frequency and the modulus of the wave vector. As a rule, the dispersion equation is cumbersome and depends on many parameters; therefore, a numerical solution of the dispersion equation or finding a neutral curve is often used. To obtain the physical consequences necessary for analyzing the conditions for the formation of cellular structures, it is necessary to use approximate formulas to obtain the dependence of the growth rate (decrement) on the modulus of the wave number vector. This model was used to study the formation of cells in the approximation of a semi-infinite layer [38–47]. Models of thermocapillary instability are currently being intensively developed in connection with the study of processes in active liquids [48].

The analysis of thermocapillary instability presents certain difficulties, since in the simplest case of a semi-infinite layer, it reduces to the analysis of a multi-parameter transcendental dispersion equation. This is characteristic of all instabilities in which viscosity is taken into account. For example, in [38], the approximation of the smallness of the Prandtl number is used ($Pr = \nu/\chi$; ν is the kinematic viscosity coefficient, χ is the thermal diffusivity coefficient) and two cases of capillary and thermocapillary waves are considered. For thermocapillary waves in the low-frequency approximation ($|\omega| << \chi k^2$, where k is the wave number, ω is the cyclic frequency), a quadratic equation is obtained for which a parametric analysis is performed. In [39], a numerical analysis of the dispersion equation for a finite layer was performed and the dependences of the increment on the wavelength were constructed. In these works, surface deformation is not fully taken into account, as will be discussed below. The influence of the boundary deformability on the thermocapillary instability of a liquid layer heated from below was studied in [41–46]. In [41, 42], the high-frequency approximation ($|\omega| >> \chi k^2$) was considered and it was assumed that the Prandtl number is of the order of unity. From the dispersion equation, we obtained the dependence of the decrement on the wave number with two maxima. For metals, the value of the Prandtl number is $Pr \sim 0.01$; therefore, the analysis of the dispersion equation proposed in these works does not fit our situation. In [43], a numerical analysis of the dispersion equation is performed taking into account surface deformation. In [44], the boundary value problem for determining the eigenvalues is numerically solved. In [45], a linear analysis of the stability of oscillatory Marangoni convection in a semi-infinite deep liquid layer with a free surface was performed. In particular, they presented some numerically calculated curves of ultimate stability and critical values of the Marangoni number for the onset of convection and the corresponding analytical results in the asymptotic limit of a high vibration frequency. In [46], the onset of Marangoni convection in a semi-infinitely deep layer of a resting liquid was investigated. An asymptotic and numerical analysis of the neutral curves of both long-wave and short-wave disturbances was carried out. The critical values of frequency and wave number are found. In [47], a thermocapillarity model was used taking into account the pressure of the incident plasma flow to explain surface-periodic structures under the influence of heterogeneous plasma flows. In [49, 50], the dispersion equation was obtained taking into account thermocapillarity for the final layer. In the shallow-depth approximation, an equation is obtained whose numerical solution made it possible to construct the dependence of the increment on the wavelength with one maximum. Thus, in most studies, the main emphasis is on obtaining and analyzing a neutral curve ($\omega = 0$). This is important information for determining the range of parameters at which instability occurs. Using this approach, the wave number is found at which the increment passes through zero, that is, the critical wave number. This turns out to be insufficient, so those wave numbers at which a maximum increment occurs play an important role. The approach based on the search for the increment maximum was successfully used in [51, 52] for the Kelvin–Helmholtz instability. In these works, for the short-wave approximation, the dispersion equation is obtained, which allows one to carry out analytical parameterization and obtain important physical consequences [51, 52], for example, the presence of two maxima

When the heat flux density exceeds 10^5 W/cm^2, an evaporation-capillary instability arises. In this case, various evaporation regimes are realized: subsonic [53] and evaporation in a vacuum or in a medium with low back pressure [54–56], and, therefore, various mechanisms of instability development. In [57], laser-induced instabilities caused by diffraction of the incident radiation at the initial roughness were considered. The interference of diffracted waves and pump waves leads to the appearance of a periodic temperature change, as a result of which there appear periodic thermocapillary forces and recoil pressure forces that increase the amplitude of the initial perturbations. Such interference instabilities lead to ordered surface relief structures with a period of the order of the wavelength and are characteristic only of pulsed laser actions. Thermocapillary flows taking into account evaporation were studied in [58–67]. In these works, the so-called long-wave approximation is used when the flow analysis is reduced to a single equation for the layer thickness.

Thus, the purpose of this section is to establish the mechanisms of behavior of the Al–Si alloy under the action of IPEB, to create a project for calculating the convective flow of molten layers in the COMSOL Multiphysics system and to analyze the data obtained, as well as to study the short-wave approximation for thermocapillary instability by analytical methods.

The experimental data on the structural-phase states of the surface layers of the alloy after the IPEB are described earlier and are based on [68, 69].

4.4.1. Methods of computer simulation in the COMSOL Multiphysics system

The creation of physical and mathematical models and calculation programs is based on COMSOL Multiphysics, which allows for the associated modelling of thermal, electromagnetic, hydrodynamic processes under the influence of heat fluxes created by electron beams.

In Fig. 4.48 is a diagram of the basic elements of a project in COMSOL Multiphysics.

• **Parameters and functions**. Assignment and description of the main parameters of the model, as well as functional dependences. These parameters and functions can be used throughout the project, for example, determined parametrically: dimensions of the model geometry; physical properties of the material of the model; impact on the model, in the form of a boundary condition.

• **Geometry**. Creating the geometry of a simulated object using geometric primitives and operations on them. The main geometric primitives for two-dimensional geometry are: point, polyline, parametric line, rectangle, ellipse, etc. To build a three-dimensional model, higher-order geometric primitives are used.

• **Material properties.** Description of thermodynamic, electromagnetic, elastic, hydrodynamic and other properties of the material in the studied model. Material properties may depend on temperature or other model parameters.

• **Physical models, initial and boundary conditions**. The main block of the model, in which the studied processes are determined and computational modules are selected for their study. The following process modules are available in COMSOL Multiphysics: heat transfer process, fluid flow, electromagnetism, elasticity and others. It is possible to use the modules together, thus obtaining a comprehensive simulation of the processes occurring in the studied model. For each module, its characteristic initial and boundary conditions are described.

• **Finite element mesh**. The type, order of finite elements is selected, and the algorithm for covering geometry with finite elements is also described.

• **Solver.** The SLAU solver is selected and configured. COMSOL Multiphysics uses an implicit method of discretization of differential equations in time – the formula for inverse differentiation (BFD). It is only possible to set the estimated time intervals, the initial and maximum time steps, as well as the minimum and maximum sampling order of BFD.

• **Processing of results**. Construction of graphs of the studied quantities. Validation of calculation results.

Based on the project structure, the computer simulation process itself in COMSOL Multiphysics is cyclical in nature and can be represented in the form of the circuit shown in Fig. 4.49, is to analyze the results and refine the parent modules with the subsequent launch of the entire project.

The whole development process can be divided into three main stages: analysis and model development; debugging release. At the first stage, an initial analysis of the studied object and processes is performed, parameters and physical models are extracted, geometry and grids are built. The stage of model debugging is cyclical and consists in the analysis of the results obtained and the refinement of higher models with the subsequent launch of the entire project. The last step is the release – this is documenting the project, formatting the display of calculation results and uploading them to files

The computer simulation technique presented in this section in the COMSOL Multiphysics system was also used to study plasma expansion in an end plasma accelerator [70]. In this work, a magnetohydrodynamic model of the flow of heterogeneous plasma during an electric explosion of conductors was developed, with the help of which the plasma current, electric current, and Lorentz force field distributions were obtained. It is shown that parallel flow in the central part of the electrode and on its periphery is unstable. A plasma focus is formed in the centre of the electrode under the action of the Lorentz force, and in the peripheral zone there are vortex structures, apparently due to the Kelvin–Helmholtz instability.

4.4.2. Mechanisms of the impact of electron beams on Al–Si alloys

Figure 4.50 presents an SEM image of the irradiated surface. From this figure it follows that the modified surface has a cellular structure. The transverse cell sizes range from 350 nm to 550 nm. Silicon particles are located at the cell boundaries, which is confirmed by the TEM analysis of the transverse section (Fig. 4.51). The presence of cells indicates high cooling rates of $\sim 10^6-10^7$ K/s. Analysis of TEM images of the cross section of the samples showed that the IPEB leads to the formation of a gradient structure

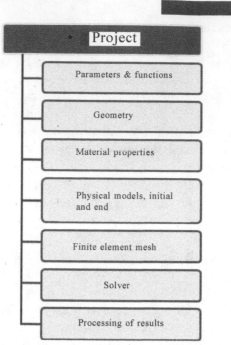

Fig. 4.48. COMSOL Multiphysics Project Tree Elements.

(Fig. 4.52).

At a depth of 0 to 100 μm from the irradiated surface, the structure of columnar crystallization is observed, and the columns are located at an angle to the direction of movement of the electron beam (Fig. 4.52 *a*). At the boundaries of the columns are interlayers of the second phases. The transverse dimensions of the columns are from 400 to 600 nm, and the interlayers of the second phases from 80 to 200 nm. The reason for the appearance of columnar crystallization is apparently the thermocapillary instability, which arises due to the presence of a temperature gradient in the liquid layer and leads to the formation of vortices (Fig. 4.53) and the displacement of particles of the second phases to the boundary of the columns [71–73].

The mechanism of this instability and the formation of vortices can be understood from the following reasoning. Consider the half-space $z > 0$, the temperature is maximum at the surface and decreases in depth. Let the liquid surface be perturbed and its analytical form $\eta(x,t)=A_0\exp(\alpha t)\cos(kx)$, where α is the exponent, A_0 is the amplitude of the perturbation of the interface, x is the coordinate, and t is time. Then, for the vertical velocity component, the expression $V_z(t,x,0) = \eta = \alpha A_0\exp(\alpha t)\cos(kx)$ follows from the kinematic boundary condition. The exponent depends on the wavenumber k. In the literature on the analysis of instabilities, the quantity α is also called the growth rate,

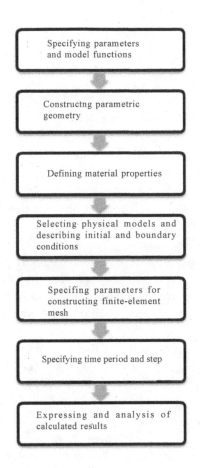

Fig. 4.49. The main stages of the project development process in COMSOL Multiphysics.

since the time derivative of $\exp(\alpha t)$ is equal to $\alpha \exp(\alpha t)$ and at $t \approx 0$ the velocity is equal to α. If $\alpha > 0$, then the amplitude of this harmonic grows exponentially, which leads to instability.

Therefore, in this case, the quantity α is called the decrement, if $\alpha < 0$, then the harmonic decays, and the quantity α is called the increment. For the horizontal velocity component, using the continuity condition, we obtain where is the amplitude of the horizontal oscillations of the melt particles. For temperature distribution, we assume that the temperature decreases with increasing z. In those areas where the temperature rises, and where the temperature drops, since cold matter is carried out from the depths. We represent the temperature perturbation in the form $T(t,x,z) = -\Theta(t,z)\cos(kx)$, where $\Theta(t,z)$ is the amplitude of the given perturbation, and the coefficient of surface tension $\sigma(t,x,z) = \sigma_0 + \sigma_T\Theta(t,z)\cos(kx)$, σ_0 is the surface tension at room temperature, and σ_T is the temperature coefficient of surface tension. For shear stress $f_x = \partial\sigma/\partial x - \sigma_T^* k(t,z)$ $\sin(kx)$. The shear stress acts parallel to the horizontal speed, so the latter increases. This is the effect of the increase in amplitude, which means instability is manifested. Such a scheme is valid under the condition $\alpha > 0$, for the opposite case ($\alpha < 0$) the system will be stable, and the vortices will twist in the opposite direction, which leads to damping. Thus, the vortex flow is present both in a stable and in an unstable case. To obtain quantitative information, it is necessary to obtain a dispersion equation to find the dependence of the decrement on the wave number (wavelength). The maximum in this dependence is important: the wavelength at which it is most unstable, and therefore it will be stored first in the form of a vortex, and then in the form of a cell.

The molten layers were also observed during the processing of the Al–Si alloy by heterogeneous plasma flows created by an electric explosion of conductors [74, 75]. However, in this case, the formed morphology of the surface layers is significantly different. When PMPJ, in the alloy, in addition to existing pores, there are additional pores associated with this exposure. Processing by a low-energy high-current electron beam leads to the fact that the pores that existed

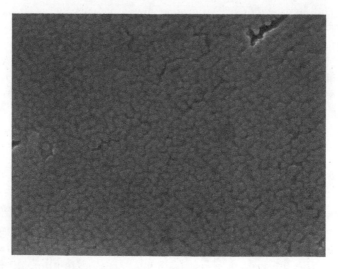

Fig. 4.50. Cell crystallization structure.

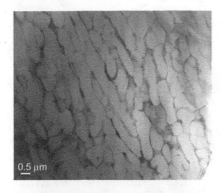

Fig. 4.51. TEM analysis of the transverse section.

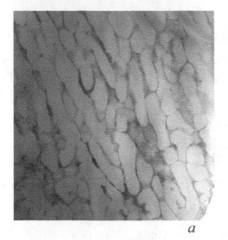

a

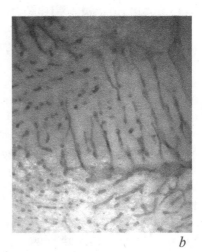

b

Fig. 4.52. TEM image of a transverse section of an Al–Si alloy at various depths subjected to IPEB; (a) the irradiated surface, b) 100 μm.

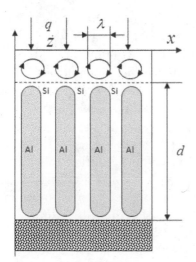

Fig. 4.53. Scheme of the impact of electron beam on an Al–Si alloy. q is the power density of the electron beam, λ is the wavelength, d is the thickness of the unmelted layer.

in the initial state are melted, and new ones do not arise due to the developed thermocapillary flow in the molten layer. The reason for the formation of various morphologies in PMPJ and IPEB is due to the fact that shock-compressed layer forms during PMPJ [76]. The existence of this layer is due to the leakage of a plasma jet, which increases the pressure. As a result, the material of the molten layer will be in the two-phase state 'liquid + bubbles'. At the end of the pulse, the shock will expand, which leads to the appearance of a rarefaction wave. This leads to expansion of the blisters. A similar effect is observed in sparkling water when opening the bottle.

At depths of 80 to 100 μm, a transition layer is observed from the crystallization zone to the heat-affected zone. The transverse dimensions of the columns at this depth are 1.0–1.2 μm. Such an increase in size can be explained by the transformation of interlayers into spherical particles (Fig. 4.52 b). Such a large thickness of the columnar crystallization layer is due to the fact that silicon and other alloying elements influence the temperature dependence of the surface tension of liquid aluminium by introducing nonlinearity [77, 78]. The nonlinear dependence of surface tension on temperature has a significant effect on the flow pattern of expanded material, which leads to the appearance of the 'heat drill' effect [79], the essence of which is the formation of a vortex structure that mixes the melt throughout its entire depth and a downward flow of fluid moving toward the centre of the bath at the cooling stage. This is manifested in an increase in the thickness of the molten layer. At a distance of 120–130 μm from the irradiated surface, primary silicon particles are detected.

As mentioned above, at a distance of 100–120 μm, silicon interlayers are transformed into spherical particles. This process is called spheroidization. According to published data [80, 81], this process has a diffusion nature, but recently, other mechanisms have been proposed in addition to diffusion. So in [82], a mechanism was proposed according to which spheroidization occurs due to the destruction of silicon wafers due to the difference in the coefficients of linear thermal expansion of the matrix and inclusion. Since, compared with the aluminium matrix, the volume fraction of silicon wafers is small, the aluminium matrix makes the main contribution to thermal expansion. The linear expansion coefficient of aluminium is 4 times that of silicon. In this regard, the thermal expansion (contraction) of the two phases is in most cases incompatible. This leads to the inevitable occurrence of mechanical stresses between the phases. Silicon inclusions are able to assimilate only 1/4 of the thermal expansion (compression) transmitted by the aluminium matrix, through its own thermal expansion (compression). The rest goes to the deformation of the matrix and the destruction of the silicon wafers (due to their brittleness). The occurrence of cracks is due to inhomogeneities of the inclusion surface. The resulting cracks will be capillaries for aluminium atoms. The mechanical stresses created by the cracks will be analogues of capillary forces that move the matrix atoms into the formed gaps between the inclusions, and in the opposite direction there are flows of vacancies and silicon atoms. The size distribution of the resulting particles is shown in Fig. 4.54. It has a bimodal character. The average particle size is 138.9 ± 45.3 nm. The distribution maxima fall on the particle sizes of 130 nm and 190 nm.

Thus, when developing mathematical models for the formation of gradient structures in IPEB, it is necessary to take into account the following factors, such as convective vortex fluid flow in the molten layer, crushing and spheroidization of silicon wafers in the heat-affected zone.

4.4.3. Thermal and thermocapillary modelling of processes occurring in Al–Si alloys under the influence of an electron beam

Simulation of the thermal effect of an electron beam

As in [51, 83], we simulate the thermal effect of an electron beam using the enthalpy approach. The advantages of this approach is that it allows you to take into account phase transitions of the first kind without involving additional conditions. We consider the effect of an electron beam with a surface energy density E_S on the sample in the form of a flat plate of thickness h. Since we are interested in the temperature distribution over the depth of the sample, we restrict ourselves to solving the one-dimensional heat conduction problem. The z axis is directed into the plate. The electron flux acts on the surface $z = 0$ for a time t_0, and there is no heat flux on the back of the plate $z = h$. The heat equation in this case will be:

$$\rho \frac{\partial H}{\partial t} = \frac{\partial}{\partial z}\left(\lambda \frac{\partial T}{\partial z}\right), \tag{1}$$

where H is the enthalpy, ρ is the density, λ is the thermal conductivity coefficient, and T is temperature. Phase transitions under the influence of an electron beam are taken into account as follows:

$$\rho H(T) = \begin{cases} \rho_S C_S T, & T < T_L \\ \rho_L \dfrac{L_L}{\Delta T_L} T, & T_L \leq T \leq T_L + \Delta T_L \\ \rho_L C_L T, & T_L + \Delta T_L \leq T < T_V \\ \rho_V \dfrac{L_V}{\Delta T_V} T, & T_V \leq T \leq T_V + \Delta T_V \\ \rho_V C_V T, & T_V + \Delta T_V \leq T \end{cases} \tag{2}$$

where C_p is the heat capacity coefficient, L is the specific heat of the phase transition, and the indices S, L, and V denote the solid, liquid, and gas phases. On the sample surface $z = 0$, the heat flux is set

$$-\lambda \frac{\partial T}{\partial z} = q_0(t), \quad q_0(t) = \begin{cases} \dfrac{E_S}{t_0} - \dot{m}(T) L_V, & 0 \leq t \leq t_0 \\ 0, & t > t_0 \end{cases} \tag{3}$$

where E_s is the electron beam energy density, $\dot{m}(T) = p_c(1-\beta)\sqrt{\dfrac{M}{2\pi RT}}$, $p_c = p_0 \exp\left(\dfrac{L_V M(T - T_V)}{RTT_V}\right)$, M is the molar mass, R is the universal gas constant, p_0 is the pressure at T_V, β is a constant.

At the boundary $z = h$:

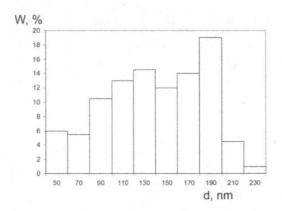

Fig. 4.54. The distribution of particles of the second phases in size (depth 100 μm).

$$\frac{\partial T}{\partial z} = 0 \qquad\qquad (4)$$

The initial temperature over the entire depth of the plate. The numerical solution of the system of equations (1)–(4) was carried out using the implicit difference scheme by the sweep method. The thermophysical constants of the Al–Si alloy were calculated according to the rule of the mixture. Table 4.12 shows the values of these constants. Since the experiment had high vacuum conditions, the value of the evaporation temperature was calculated according to the Clapeyron–Clausius equation. It amounted to 1270 K. The calculations were performed for electron beams with an energy density varying in the range from 15 to 35 J/cm^2 and a pulse duration of $t_0 = 150$ μs.

Figure 4.55 shows the temperature distribution over time at various distances from the irradiated surface when exposed to an electron beam with an energy density of 35 J/cm^2. It can be seen that in graphs 1–6 after the end of the pulse until 300 μs, the cooling rate of the substance is negligible. This indicates the presence of a vapour layer that holds high temperatures on the surface. At $t > 300$ μs, the cooling rate increases sharply, and its values on the surface and at depths up to 20 μm (curves 1–3) are higher than at depths from 20 to 50 μm (curves 4–6). This indicates the expansion of the vapor layer. An estimate of the cooling rate of the surface layers showed that it reaches ~10^6 K/s. At such values of the cooling rate, a cellular crystallization structure is formed. At a depth of 80 μm (curve 7), the temperature is less than the eutectic temperature. At such depths, spheroidization of silicon particles predominates [82].

An analysis of the dependence of temperature on the coordinate (Fig. 4.56) showed that by the time moment of 50 μs the material is still in the solid state (curve 1). In this case, the action of the electron beam also reduces to the generation of a thermoelastic wave [51]. At $t = 100$ μs (curve 2), the material in the range from 0 to 10 μm is in the molten state. Here, the convective flow of the melt begins to develop. By the time the pulse ends (curve 3) in the range from 0 to 15 μm, the substance is in a gaseous state, while the thickness of the evaporated layer increases up to a time of 300 μs (curve 4). Then a gradual temperature equalization in depth is observed (curves 5 and 6). Calculations of the penetration depth

(Table 4.13) showed that the thickness of the molten layer increases with increasing energy density of the electron beam. Its values practically coincide with the experimental values [84],

The obtained data on the penetration depth and surface temperature will be used to estimate the temperature gradient when creating a thermocapillary model for the formation of surface cellular structures.

Thermocapillary model

To solve the problem, we choose a Cartesian coordinate system (x, y, z) and consider a viscous incompressible heat-conducting fluid that occupies a layer of thickness h on a free surface $z = \eta(x,y,z)$ and absorbs heat. For heat flux values of $\sim 10^5$ W/cm^2 used in the experiment, the approximation of an incompressible fluid should be considered justified [85]. After exposure to an electron beam, a temperature profile $T = T_0(z)$ is established in the liquid layer. Let the perturbation wave vector be directed in the XOY plane. Then they depend on the x, y coordinates and time according to the law $\exp(\omega t - i(mx + ly))$, where i is the imaginary unit, m and l are the projections of the wave vector on the X and Y axis, respectively.

To study the instability of a stationary state, we linearize the Navier–Stokes equations and the heat equation for temperature. They will have the form:

Table 4.12. Thermophysical characteristics of an Al–Si alloy [163]

Symbol/dimension	Al–Si alloy	Physical value
T_L, K	850	Eutectic temperature
T_V, K	1270	Evaporation temperature
ρ_s, kg/m^3	2656	Density in solid state
ρ_L, kg/m^3	2398	Density in liquid state
C_s, J/(kg K)	880	Specific heat in solid state
C_L, J/(kg K)	1160	Specfic heat in liquid state
λ_s, W/(m K)	200	Heat conductivity in solid state
λ_L, W/(m K)	86	Heat conductivity in liquid state
L_L, kJ/kg	385	Specific heat of melting
L_V, kJ/kg	10444	Specific heat of evaporation

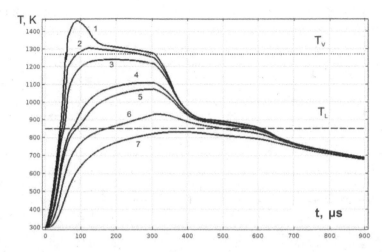

Fig. 4.55. The dependence of temperature on time at various distances from the irradiated surface 1) 0 μm, 2) 10 μm, 3) 20 μm; 4) 20 μm; 5) 30 μm; 6) 50 μm; 7) 80 μm.

$$\frac{\partial u}{\partial \tilde{n}}\, x = -\frac{\partial p}{\partial} + i\left(\frac{\partial^2 u}{\partial x^2} + \frac{\partial^2 u}{\partial y^2} + \frac{\partial^2 u}{\partial z^2}\right),\ \frac{\partial v}{\partial \tilde{n}}\, y = -\frac{\partial p}{\partial} + i\left(\frac{\partial^2 v}{\partial x^2} + \frac{\partial^2 v}{\partial y^2} + \frac{\partial^2 v}{\partial z^2}\right),$$

$$\frac{\partial w}{\partial \tilde{n}}\, z = -\frac{\partial p}{\partial} + i\left(\frac{\partial^2 w}{\partial x^2} + \frac{\partial^2 w}{\partial y^2} + \frac{\partial^2 w}{\partial z^2}\right),\ \frac{\partial u}{\partial y} + \frac{\partial v}{\partial x} + \frac{\partial w}{\partial} = 0,$$

$$\frac{\partial T}{\partial t} + w\frac{d T_0}{d z} = \left(\frac{\partial^2 T}{\partial x^2} + \frac{\partial^2 T}{\partial y^2} + \frac{\partial^2 T}{\partial z^2}\right) \tag{5}$$

where u, v, w are the components of the perturbed velocity vector, c is the melt density, p, T are the pressure and temperature perturbations, are the kinematic viscosity, thermal diffusivity, and surface tension, respectively. We believe that surface tension depends on temperature according to a linear law

$$\sigma = \sigma_0 + \sigma_T(T - T_0) \tag{6}$$

where σ_T is the temperature coefficient of surface tension, σ_0 is the surface tension at room temperature T_0.

The boundary conditions on the surface of the melt $z = 0$ have the form:

Table 4.13. Dependence of the penetration depth of an Al–Si alloy and aluminium on the surface energy density of a beam

E_s, J/cm²	Al–Si alloy d, mm	Aluminium d, μm	Experimental values d, μm
15	23	18	23
20	39	39	30
25	54	57	55
35	80		80

$$-p + 2v\rho\frac{\partial w}{\partial z} = \sigma\left(\frac{\partial^2\eta}{\partial x^2} + \frac{\partial^2\eta}{\partial y^2}\right), \ \rho v\left(\frac{\partial u}{\partial z} + \frac{\partial w}{\partial x}\right) = \frac{\partial\sigma}{\partial x}, \ \rho v\left(\frac{\partial v}{\partial z} + \frac{\partial w}{\partial y}\right) = \frac{\partial\sigma}{\partial y}$$

$$\frac{\partial T}{\partial z} = 0, \ \frac{\partial\eta}{\partial t} = w. \tag{7}$$

At the melt–solid boundary $z = -h$, the conditions of adhesion and impermeability act:

$$u = v = w = 0. \ T = 0, \tag{8}$$

The gradient of surface tension along the x and y axes has the form:

$$\frac{\partial\sigma}{\partial x} = \frac{\partial\sigma}{\partial T}\left(\frac{\partial T}{\partial x} + a\frac{\partial T_0}{\partial z}\eta_x\right), \ \frac{\partial\sigma}{\partial y} = \frac{\partial\sigma}{\partial T}\left(\frac{\partial T}{\partial y} + a\frac{\partial T_0}{\partial z}\eta_y\right) \tag{9}$$

Here, parameter a was introduced in order to compare the dispersion equations obtained for $a = 0$ and $a = 1$. If we put $\sigma_0 = 0$ and $a = 0$,

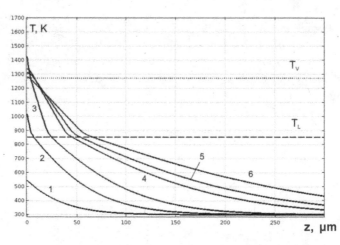

Fig. 4.56. The dependence of temperature on the distance from the irradiated surface at various points in time (1) 50 μs, 2) 100 μs, 3) 150 μs, 4) 300 μs, 5) 400 μs. 6) 600 μs).

$\sigma T \neq 0$, then we obtain the problem of thermocapillary instability with an undeformed plane boundary (Pearson problem) . We seek the solution of equations (5) in the form

$$u(x,y,z,t)=U(z)\exp(\omega t - i(mx-ly)), \quad v(x,y,z,t)=V(z)\exp(\omega t - i)(mx+ly)$$
$$w(x,y,z,t)=W(z)\exp(\omega t - i(mx+ly)),$$
$$T(x,y,z,t)=T(z)\exp(\omega t - i(mx+ly)), \eta(x,y,t)=\eta_0\exp(\omega t - i(mx+ly)),$$

$$(10)$$

where $U(z), V(z), W(z), P(z), T(z), \eta_0$ are the amplitudes of the perturbations of the projections of the velocity vector on the axis, pressure, temperature and surface, respectively; $k = (l, m)$ is the wave vector, i is the imaginary unit. Substitution of (10) into (6) - (8) leads to the following system

$$P(z)=i\frac{\rho\nu}{k^2}\left(W'''(z)-k_1^2 W'(z)\right), \quad W''(z)-k_1^2 W'(z)-\frac{1}{\rho\nu}P'(z)=0,$$

$$T''(z)-k_2^2 T(z)-G_0 W(z)/\chi=0.$$

$$(11)$$

where $k_1^2=\dfrac{\omega}{\nu}+k^2$, $k_2^2=\dfrac{\omega}{\chi}+k^2$, $G_0=\dfrac{dT_0}{dz}$. After the conversion, the first and second equations of system (11) take the form:

$$W^{IV}(z)-(k^2+k_1^2)W''(z)+k^2 k_1^2 W(z)=0$$
$$T''(z)-k_2^2 T(z)-\frac{G_0}{\chi}W(z)=0$$

$$(12)$$

The boundary conditions will be as follows

$$-P(0)+2\rho\nu W'(0)=-\sigma_0 k^2 \eta_0, \quad \rho\nu\left(U'(0)-imW(0)\right)=-i\sigma_T G_0 m \Sigma_0$$
$$\rho\nu\left(V'(0)-ilW(0)\right)=-il\sigma_T G_0 \Sigma_0$$
$$T'(0)=0, \quad \omega\eta_0 = W(0), \quad \Sigma_0 = a\eta_0 + T(0)/G_0$$
$$W'(-h)=W(-h)=T(-h)=0$$

$$(13)$$

After transformations (13) takes the form

$$W'''(0)-(k_1^2+2k^2)W'(0)-k^3\omega_c^2 W(0)/(\omega\omega_\nu)=0,$$
$$W''(0)+k^2 W(0)+\omega_T k^2 \Sigma_1(0)/\omega = 0.$$
$$T'(0)=0, \quad W'(-h)=W(-h)=T(-h)=0$$

$$(14)$$

where $\omega_c^2 = \dfrac{\sigma_0}{\rho} k^3$, $\omega_T = \dfrac{\sigma_T G_0}{\nu \rho}$, $\omega_\nu = \nu k^2$, $\Sigma_1(0) = aW(0) + (\omega / G_0)T(0)$.

We assume that $h \to -\infty$. Solution (12) that satisfies the boundary conditions on $z \to -\infty$ has the form:

$W(z) = A_1 \exp(kz) + A_2 \exp(k_1 z)$,

$$T(z) = C \exp(k_2 z) - \dfrac{G_0}{\omega} \left(A_1 \exp(kz) - A_2 \delta \exp(k_1 z) \right); \delta = \dfrac{\nu}{\chi - \nu}. \tag{15}$$

We define $\Sigma_1(0)$ in terms of the coefficients A_1 and A_2

$$\Sigma_1(0) = aW(0) + \omega / G_0 T(0) = a(A_1 + A_2) + \dfrac{1}{k_2}\left(A_1(k - k_2) + \delta A_2(k_2 - k_1) \right) =$$

$$= (a - 1 + k / k_2)A_1 + (a + \delta - \delta k_1 / k_2)A_2. \tag{16}$$

Substituting (15) and (16) into the first and second equations (14) allows the system of equations for A_1 and A_2

$$\left(\omega \omega_\nu (z_1^2 + 1) + \omega_c^2 \right) A_1 + \left(2z_1 \omega \omega_\nu + \omega_c^2 \right) A_2 = 0,$$

$$\left(\omega_T (z_2(a-1)+1) - 2z_2\omega \right) A_1 + \left(\omega_T ((az_2 + \delta(z_2 - z_1) - z_2\omega(z_1^2 + 1)) \right) A_2 = 0;$$

$$z_1 = \dfrac{k_1}{k} = \sqrt{1 + \omega / \nu k^2}, z_2 = \dfrac{k_2}{k} = \sqrt{1 + \omega / \chi k^2}, \delta = \dfrac{\varepsilon}{1 - \varepsilon}, \varepsilon = \dfrac{\nu}{\chi} \tag{17}$$

To search for a nonzero solution to system (17), we require that the determinant be zero. As a result, we obtain an equation that relates the frequency (ω) and the modulus of the wave vector (k), that is, it is a dispersion equation. We represent this equation in the form where

$$R_T = \omega_\nu \omega_T \begin{pmatrix} (1 - z_1 / z_2)\delta \left(\omega^2 + 2\omega\omega_\nu + \omega_c^2 \right) + \\ + (1 - 1/z_2)\left(2\omega\omega_\nu z_1 + \omega_c^2 \right) + a\omega\left(\omega + 2\omega_\nu(1 - z_1) \right) \end{pmatrix}$$

$$R_\sigma = \omega^2 \left((\omega + 2\omega_\nu)^2 + \omega_c^2 - 4z_1\omega_\nu^2 \right). \tag{18}$$

From the obtained relations for $a = 0$, we obtain the dispersion equation [38]:

$$\omega^2\left((\omega+2\omega_v)^2+\omega_c^2-4z_1\omega_v^2\right)-$$

$$-\omega_v\omega_T\left((1-z_1/z_2)\delta\left(\omega^2+2\omega\omega_v+\omega_c^2\right)+(1-1/z_2)\left(2\omega\omega_vz_1+\omega_c^2\right)\right)=0 \quad (19)$$

For $a=1$ equation [21]:

$$\omega^2\left((\omega+2\omega_v)^2+\omega_c^2-4z_1\omega_v^2\right)-$$

$$-\omega_v\omega_T\left((1-z_1/z_2)\delta\left(\omega^2+2\omega\omega_v+\omega_c^2\right)+(1-1/z_2)\left(2\omega\omega_vz_1+\omega_c^2\right)+\omega(\omega+2\omega_v(1-z_2))\right).$$

$$(20)$$

Consider the low-frequency mode, for which the inequality $|\omega|\ll \chi k^2$ holds, in addition, for liquid metals $\varepsilon\ll1$. Therefore, we can write approximate relations, neglecting:

$$\frac{1}{z_2}\approx1-\frac{\varepsilon\omega}{2\omega_v}\Rightarrow1-1/z_2\approx\varepsilon\omega/2\omega_v \text{ and } (1-z_1/z_2)\delta\approx(1-z+z\varepsilon\omega/2\omega_v)\varepsilon\approx(1-z_1)\varepsilon$$

$$(21)$$

Taking into account (21), equations (18) will take the form:

$$R_T=\frac{\omega_T\varepsilon}{2}(\omega+2\omega_v(1-z_1))\left(2\omega\omega_v\left(1+\frac{a}{\varepsilon}\right)+\omega_c^2\right)$$

$$R_\sigma=\omega^2\left((\omega+2\omega_v)^2+\omega_c^2-4z_1\omega_v^2\right) \quad (22)$$

We use a replacement. As a result, we get

$$R_T=\frac{\omega_T\omega_v}{2}(z_1-1)^2\left(2(z_1^2-1)(a+\varepsilon)\omega_v^2+\varepsilon\omega_c^2\right)$$

$$R_\sigma=\omega_v^2(z_1^2-1)^2\left(\omega_v^2(z_1^2-1)^2+\omega_c^2-4z_1\omega_v^2\right). \quad (23)$$

Then the dispersion equation takes the form

$$(z_1+1)^2\left(\left(z_1^2+1\right)^2+C^2=4z_1\right)-B\left(2\left(z_1^2-1\right)(a+\varepsilon)+\varepsilon\omega_c^2/\omega_v^2\right)=0 \quad (24)$$

where $C^2=\dfrac{\omega_c^2}{\omega_v^2}$, $B=\dfrac{\omega_T}{2\omega_v}$. Unstable will be solutions that satisfy the condition Re $\omega>0$ and Re $z1<0$. The gradient of the unperturbed temperature is defined as $G_0=\dfrac{T-T_L}{h}$, where T is the maximum temperature, T_L is the eutectic temperature. For $E_s=35$ J/cm^2, according to Table 4.13, the thickness is $h = 80$ μm, then $G_0 = 7.25 \cdot 10^6$ K/m. The numerical solution of equation (24) at $a = 0$ shows that waves with $\lambda > 60$ μm are unstable (Fig. 4.57, curve 1). The maximum instability occurs at a wavelength of 261 μm. At $a = 1$ (Fig. 4.57, curve 2), the instability will begin with $\lambda > 65$ μm, and its maximum will fall at 254 μm. Such wavelengths correspond to the sizes of surface periodic structures observed in experiments on electric explosive [46] and laser [86] surface treatments. At the same time, the sizes of the cellular structures have sizes from 350 nm to 550 nm (see the previous paragraph). Such values are possible at a temperature gradient of $G_0 \sim 10^{10}$–10^{12} K/m. Solution (24) at a value of $G_0 \sim 10^{10}$ K/m shows that at $a = 0$ the instability will begin with $\lambda > 0.8$ μm, and its maximum will fall at 1.8 μm (Fig. 5.58 a).

If we put $a = 1$, then the onset of instability will occur at $\lambda = 16$ μm, and the maximum at a wavelength of 35 μm (Fig. 4.58 b). This means that the low-frequency approximation does not provide adequate agreement with experimental data on the sizes of crystallization cells. Therefore, we will analyze equation (19). Using the substitution and subsequent transformations, this equation is reduced to an algebraic equation of the 16th degree, which, due to its cumbersome nature, will not be written out. In this case, solutions that satisfy the condition and will be unstable.

The results of a numerical solution with a temperature gradient of $G_0 = 2.9 \cdot 10^{10}$ K/m are presented in Fig. 4.59. From this figure it follows that the first two solutions of equation (19) are unstable

This suggests the existence of two dependences of the decrement on the wavelength. In the first case, the decrement maximum occurs at a wavelength of $\lambda = 140$ nm, with $\alpha_m \approx 10^9$ s^{-1} (Fig. 4.59 a). Comparison with the experimental SEM data showed that its values are 3-4 times smaller than the cell sizes. Such a discrepancy can be explained by the fact that in this work we did not take into account the effect of silicon on the surface tension and its temperature dependence. Taking this influence into account is the task of a separate article. On the other hand, the dependence of α on λ shown in Fig. 4.59 a, is single-mode. The size distribution of high-speed crystallization cells is of the same nature. This allows us to conclude that thermocapillary instability adequately explains the single-mode nature of this distribution. In the second case, two peaks are observed at wavelengths $\lambda = 2.6$ μm and $\lambda = 9.5$ μm (Fig. 4.59 b), with $\alpha m \approx 10^6$ s^{-1}. With a decrease in the temperature gradient, both in the first and second cases, the wavelengths at which the maximum growth rate falls are increased. The presence of two dependences of decrement on the wavelength allows us to make the assumption that the first dependence describes the appearance of crystallization cells on the surface, and the second describes the formation of columnar structures at a depth of 80 μm.

Numerical simulation of thermocapillary flows

Figure 4.60 shows the results of numerical simulation of thermocapillary convection for the case α < 0. As mentioned above, in this case a vortex flow will also be observed in the molten layer, however, the vortices will twist in the opposite direction. As a result, surface disturbances decay. Similar vortex flow patterns were observed in [87, 88] when modeling the convective instability of multilayer systems in the presence of a temperature gradient.

Thus, from the analysis of experimental data it was revealed that the surface layer is represented by a columnar crystallization structure with interlayers of the second phases, the transverse dimensions of which depend on the distance from the irradiated surface and range from 400 to 600 nm, and interlayers of the second phases from 80 to 200 nm in the surface layers and at depths from 80 to 100 μm - from 1 to 2 μm.

Projects have been created in the COMSOL Multiphysics system of the action of an electron beam on the surface layers of an Al–Si alloy, including thermal exposure and thermocapillary flow.

Calculations of the penetration depth in the created COMSOL Multiphysics project showed that the thickness of the molten layer increases with increasing electron beam energy density and its values coincide with experimental data. The obtained data on the penetration depth and surface temperature were used to estimate the temperature gradient when creating a thermocapillary model for the formation of surface cellular structures.

A mechanism is proposed for the formation of a cellular and columnar crystallization structure, which consists in the occurrence of thermocapillary instability of the vacuum – molten metal interface. An analysis of the initial stage of the development of thermocapillary instability by solving the dispersion equation for thermocapillary waves showed that the low-frequency approximation provides an adequate explanation of the appearance of a nanoscale columnar structure at a temperature gradient of $G_0 \sim 10^{10}-10^{12}$ K/m. The solution of the complete dispersion equation shows that there are two dependences of the decrement on the wavelength. The first

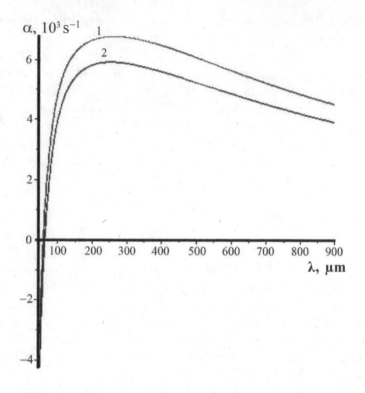

1) $a = 0$, 2) $a = 1$

Fig. 4.57. Dependence on the wavelength in the low-frequency approximation (24) at $G_0 = 7.25 \cdot 10^6$ K/m.

dependence has one maximum in the nanoscale range, which allows us to conclude that it is responsible for the formation of a cellular structure on the surface. The second dependence has maxima in the microsize range.

The results of numerical modelling of thermocapillary convection, in which a multi-vortex flow is observed in a stable mode, are presented.

4.5. Conclusions for chapter 4

1. The surface layer of the Al–Si alloy was modified using an electron beam of various energy densities (from 10 to 35 J/cm²);

2. A set of tests was carried out for physical and mechanical properties (microhardness, nanohardness, friction coefficient, wear resistance, etc.) of samples after energy exposure in various modes to establish their optimal parameters;

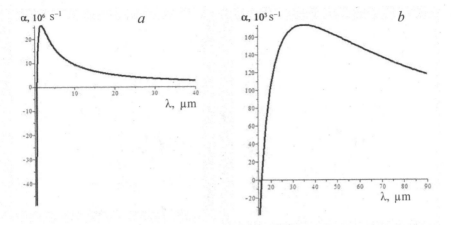

Fig. 4.58. Dependences on the wavelength in the low-frequency approximation (24) at $G_0 = 5.8 \cdot 10^{10}$ K /m; *a*) $a = 0$, b) $a = 1$

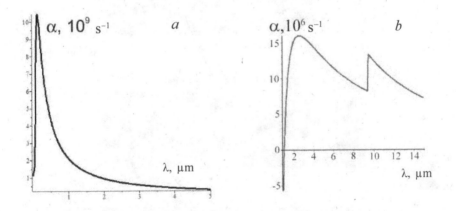

Fig. 4.59. Dependence of the decrement on the wavelength at G0 = $2.9 \cdot 10^{10}$ K/m obtained by solving equation (19). *a*) the first root of equation (19), *b*) the second root of equation (19).

3. It was shown that the value of microhardness after treatment with an intense electron beam depends on the beam energy density and reaches maximum values at processing parameters of 25, 30 and 35 J/cm² (0.93 ± 0.052 GPa for 25 J/cm², 0.97 ± 0.071 GPa for 30 J/cm², 0.96 ± 0.103 GPa for 35 J/cm²). The presented modes are defined as optimal;

4. As a result of tribotechnical tests, it was found that simultaneously with an increase in microhardness in irradiated

162

samples, a decrease in the friction coefficient and wear rate is observed. Compared to the material in the delivery state, the friction coefficient decreased by ≈1.3 times, the wear rate by ≈6.6 times.

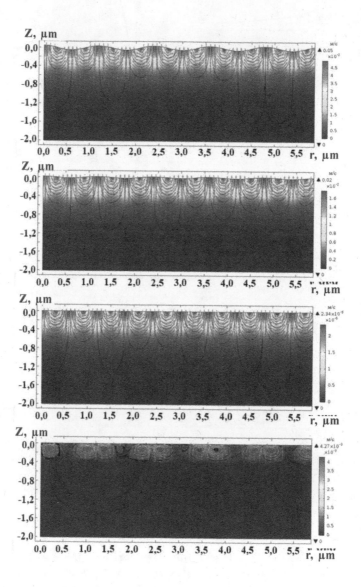

Fig. 4.60. The velocity field of the flow of the molten layer at different points in time for α <0. *a*) 1 μs, *b*) 2 μs, *c*) 10 μs, *d*) 12 μs.

5. Using atomic force microscopy, it has been established that a fine-grained cellular structure is formed in the treated layer, and there are no defects in the form of micropores. The roughness of the treated layer of Al–Si alloy samples is in the range from 17 to 90 nm, and the substrate near the treated layer is from 30 to 77 nm.

6. It was established that treatment with intense electron beams leads to homogenization of the surface layers of the Al–Si alloy. The thickness of the homogenized layer varies depending on the IPEB parameters and reaches maximum values of 100 μm at an energy density of 35 J/cm^2.

7. It was found that the modified layer, from 35 μm to 100 μm thick, depending on the electron beam energy density, is free from intermetalides and consists of a nanocrystalline structure of cellular crystallization. It has been suggested that these two factors are responsible for the increased mechanical characteristics of the modified layer.

8. It was determined that the melting of an Al–Si alloy by an intense pulsed electron beam is accompanied by a decrease in the concentration of silicon in the surface layer with a thickness of up to 30 μm.

9. It was revealed that irradiation is accompanied by the formation of a gradient structure – the structure of cellular crystallization passes, with distance from the processing surface, into a mixed type structure, in which, along with cells, partially dissolved cast inclusions are present.

10. A mechanism is proposed for the formation of a cellular and columnar crystallization structure, which consists in the occurrence of thermocapillary instability of the vacuum–molten metal interface. An analysis of the initial stage of the development of thermocapillary instability, by solving the dispersion equation for thermocapillary waves, showed that the low-frequency approximation provides an adequate explanation for the appearance of a nanoscale columnar structure at a temperature gradient of $G_0 \sim 10^{10}$–10^{12} K/m.

11. Using the COMSOL Multiphysics system, a mathematical model of the thermal effect of an electron beam on the surface layers of an Al–Si alloy has been developed. Calculations of the penetration depth in the created COMSOL Multiphysics project showed that the thickness of the molten layer increases with increasing electron beam energy density and its values coincide with experimental data. The obtained data on the penetration depth and surface temperature

were used to estimate the temperature gradient when creating a thermocapillary model for the formation of surface cellular structures. The results presented in the chapter are published in [89–103].

References for Chapter 4

1. Milman, Yu.V. Plasticity characteristic obtained by indentation. Journal of Physics D: Applied Physics, 2008. 41 (7), article No. 074013
2. Laskovnev, A.P. Modification of the structure and properties of eutectic silumin by electron-ion-plasma treatment / A. P. Laskovnev, Yu.F. Ivanov, E.A. Petrikova [et al.]. - Minsk: Belarus. Navuka, 2013 .
3. Panin, S.V., Vlasov, I.V., Sergeev, V.P., Ovechkin, B.B., Lyubutin, P.S., Ramasubbu, S., Mironov, Y.P., Maruschak, P.O. Effect of vacuum arc ion beam treatment on the structure and mechanical properties of 30CrMnSiNi2A steel. Physical Mesomechanics, 2016, 19 (4), pp. 392–406.
4. Zhang, C., Lv, P., Xia, H., Yang, Z., Konovalov, S., Chen, X., Guan, Q. The microstructure and properties of nanostructured Cr–Al alloying layer fabricated by high-current pulsed electron beam. Vacuum, 2019. V. 167, pp. 263–270.
5. Rotshtein, V. Surface treatment of materials with low-energy, high-current electron beams / "Materials surface processing by directed energy techniques" / V. Rotshtein, Yu. Ivanov, A. Markov; Ed. by Y. Pauleau. Ch. 6. - Elsevier, 2006. - P.205–240.
6. Ivanov, Yu. Evolution of Al–19.4 Si alloy surface structure after electron beam treatment and high cycle fatigue / Yu. F. Ivanov, K. V. Alsaraeva, V. E. Gromov, S. V. Konovalov, O. A. Semina // Materials Science and Technology (United Kingdom). – 2015. - No. 31 (13a). - P. 1523–1529.
7. Bykov, I.V. Method for point-by-point measurements of the relief, interaction forces, and local properties: a new approach for complex analysis in atomic force microscopy / I.V. Bykov // Scientific Instrument Making. - 2009. - No. 4. - V. 19. - P. 38–43.
8. Barannikova, S.A. The use of atomic force microscopy methods to study the structure of steel 40Kh13 after tempering / S.A. Barannikova, G.V. Shlyakhova, L.B. Zuev // Vestn. Tambov. Univ.. Ser .: Natural and technical sciences. - 2016. - No. 3. - S. 882–885.
9. Shlyakhova, G.V., Bochkareva, A.V., Barannikova, S.A., Zuev, L.B., Martusevich, E.V. Application of atomic force microscopy for stainless steel microstructure study at various kinds of heat treatment. Izv. Ferrous Metallurgy, 2017. 60 (2), pp. 133–139.
10. Shlyakhova, G.V., Bochkareva, A.V., Barannikova, S.A., Zuev, L.B., Martusevich, E.V. Microstructure of stainless steel after heat treatment: Data from atomic-force microscopy. Steel in Translation, 2017.47 (2), pp. 99–104.
11. Ulyanov, P.G. The use of a microscope of atomic forces to study the substructure of grains of structural steels / P.G. Ulyanov, D.Yu. Usachev, A.M. Dobrotvorsky et al. // Vestn. St. Petersburg. Univ.. Ser. 4: Physics. Chemistry. - 2010. - No. 4. - P. 44–48.
12. Shur, V.Ya. The formation of nanodomain structures as a result of pulsed laser irradiation

of lithium niobate / V.Ya. Shur, D.K. Kuznetsov, A.I. Lobov et al. // Izv. RAS. Ser. physical. - 2008. - No. 72. - Issue. 2. - P. 198–200.

13. Ferraris, S. Surface structuring by Electron Beam for improved soft tissues adhesion and reduced bacterial contamination on Ti-grade 2 / S. Ferraris, F. Warchomicka, C. Ramskogler et. al.// J. of Mat. Proc. Tech. - 2019 . V. 266. - P. 518–529.

14. Osintsev, K.A. AFM investigation of silumin structure modified by Al–Y2O3 coating using the method of electric explosive alloying / K. A. Osintsev, K. A. Butakova, S. V. Konovalov, D. V. Zagulyaev, V. E. Gromov, // IOP Conference Series: Materials Science and Engineering. - 2018 .- Vol. 411. - P. 1–8.

15. Osintsev, K.A. The study on mechanical properties of Al–Y_2O_3 coatings formed on silumin by electroexplosive spraying / Osintsev K.A., Shlyarov V.V., Butakova K.A., Zagulyaev D.V., Konovalov S.V., V.E. Gromov // The 9th International Symposium on Materials in External Fields (ISMEF 2018). - 2018 .- P. 24.

16. Belov, N.A. Phase composition and structure of silumins / N.A. Belov, S.V. Savchenko, A.V. Hwan. - M.: MISIS, 2008 .282 p.

17. Zolotorevsky V.S. Metallurgy of foundry aluminium alloys / V.S. Zolotorevsky, N.A. Belov. - M.: MISiS, 2005 . 376 p.

18. Belov N.A. The phase composition of aluminium alloys / N.A. Belov. - M .: Publishing House MISiS, 2009. - 392 p.

19. Physical quantities. Reference book / A.P. Babichev et al .; under the general. ed. I.S. Grigoryeva, E.Z. Meilikhova. - M .: Energoatomizdat, 1991 .1232 p.

20. Koval ', N. N. A facility for metal surface treatment with an electron beam / N.N. Koval', P.M. Shchanin, V. N. Devyatkov, V.S. Tolkachev, L.G. Vintizenko // Instruments and Experimental Techniques. - 2005. - V. 48. - P. 117 – 121.

21. Vorobyov, M. S. An electron source with a multiaperture plasma emitter and beam extraction into the atmosphere / M.S. Vorobyov, N.N. Koval, S.A. Sulakshin // Instruments and Experimental Techniques. - 2015. - V.58. - P. 687–695.

22. Koval ', N.N. Nanostructuring of surfaces of metalloceramic and ceramic materials by electron-beams / N. N. Koval, Yu. F. Ivanov // Russian Physics Journal. - 2008 - V. 51. - P. 505 – 516.

24. Fetzer, R. Electron beam facility GESA–SOFIE for in-situ characterization of cathode plasma dynamics / R. Fetzer, W. An, A. Weisenburger, G. Müller // Vacuum. - 2017. - V. 145. - P. 179 – 185.

25. Shulov, V.A. Application of high-current pulsed electron beams for modifying the surface of gas-turbine engine blades / V.A. Shulov, A.N. Gromov, D.A. Teryaev, V.I. Engel'ko // Russian Journal of Non-Ferrous Metals. - 2016. - V. 57. - P. 256 – 265

26. Proskurovsky, D. I. Use of low-energy, high-current electron beams for surface treatment of materials / D.I. Proskurovsky, V.P. Rotshtein, G.E. Ozur // Surface Coat. Technol. - 1997. –V. 96.– P. 117 – 122.

27. Proskurovsky, D. I. Pulsed electron-beam technology for surface modification of metallic materials / D. I. Proskurovsky, V. P. Rotshtein, G.E. Ozur, A. B. Markov, D. S. Nazarov, M. A. Shulov, Yu. F. Ivanov, R. G. Buchheit // Journal of Vacuum Science and Technology A: Vacuum, Surfaces and Films. - 1998 - V.16 - P. 2480 – 2488.

28. Ozur, G.E. Generation of Low-Energy High-Current Electron Beams in Plasma-Anode Electron Guns / G.E. Ozur, D.I. Proskurovsky // Plasma Physics Reports. - 2018. - V. 44. - P. 18 – 39.

29. Diankun, Lu High-Current Pulsed Electron Treatment of Hypoeutectic Al – 10 Si Alloy / Lu Diankun, Gao Bo, Zhu Guanglin, Lv Jike, Hu Liang // High Temperature Materials and Processes. - 2017 - V. 36. - P. 97 – 100.

30. Hao, Y. Surface modification of Al–12.6 Si alloy by high current pulsed electron beam

/ Y. Hao, B. Gao, G. F. Tu et al. // Applied Surface Science. - 2012. - V. 258. - P. 2052 – 2056.

31. Gao, B. Study on the nanostructure formation mechanism of hypereutectic Al – 17.5 Si alloy induced by high current pulsed electron beam / B. Gao, L. Hu, S. Li, Y. Hao, Y. Zhang, G. Tu // Applied Surface Science. - 2015. - V. 346. - P. 147 – 157.

32. Rygina, M. E. Modification of the sample's surface of hypereutectic silumin by pulsed electron beam / M. E. Rygina, Yu. F. Ivanov, A. P. Lasconev, A. D. Teresov, N. N. Cherenda, V. V. Uglov, E. A. Petrikova, M. V. Astashinskiy // IOP Conf. Series: Materials Science and Engineering. –2016. –V. 124. - P. 012138.

33. Gromov, V. E. Structural Evolution of Silumin Treated with a High Intensity Pulse Electron Beam and Subsequent Fatigue Loading up to Failure / V.E. Gromov, Yu.F. Ivanov, A.M. Glezer, S.V. Konovalov, K.V. Alsaraeva // Bulletin of the Russian Academy of Sciences. Physics - 2015 .- V. 79. - P. 1169 – 1172.

34. Feng, Jicai Microstructure evolution of electron beam welded Ti3Al – Nb joint / Jicai Feng, Huiqiang Wu, Jingshan He, Bingang Zhang // Materials Characterization. - 2005. - V. 54 - P. 99 – 105.

35. Biamino, S Electron beam melting of Ti – 48Al – 2Cr – 2Nb alloy: Microstructure and mechanical properties investigation / S. Biamino, A. Penna, U. Ackelid, S. Sabbadini, O. Tassa et al // Intermetallics. - 2011. - V. 19. - P. 776 – 781.

36. Rotshtein, V. P. Surface modification and alloying of metallic materials with low-energy high-current electron beams / V.P. Rotshtein, D.I. Proskurovsky, G.E. Ozur, Yu.F. Ivanov, A.B. Markov // Surface and Coatings Technology. - 2004. - V. 180–181. - P. 377 – 381.

37. Mirzoev, F. Kh. Laser control processes in solids / F.Kh. Mirzoev, V.Ya. Panchenko, L.A. Shelepin // Phys. Usp. - 1996. - V. 39. - P. 1 – 29.

38. Bugaev, A. A. Thermocapillary phenomena and the formation of surface relief under the influence of picosecond laser pulses / A.A. Bugaev, V.A. Lukoshkin, V.A. Urpin, D.G. Yakovlev // Journal of Technical Physics. - 1988 .-- V.58. - No. 5. - P. 908 – 914.

39. Urpin, V.A. Excitation of capillary waves in inhomogeneously heated liquid films / V.A. Urpin, D.G. Yakovlev // Journal of Technical Physics. - 1989 .- V.59. - No. 6. - S. 19 – 25.

40. Levchenko, E. B. The instability of surface waves in the inhomogeneously heated liquid / E.B. Levchenko, A.L. Chernyakov // Sov. Phys .- JETP. - 1981. - V. 54. - P. 102 – 105.

41. Levchenko, E. B. Instability of capillary waves in an inhomogeneously heated liquid under the influence of laser radiation / E.B. Levchenko, A.L. Chernyakov // Fizika I Khimiya Obrabotki Materialov.– 1983. - No. 1. - P. 129 – 141. [in Russian]

42. Takashima, M. Surface tension driven instability in a horizontal liquid layer with a deformable free surface / M. Takashima // J. Phys. Soc. Japan - 1981. - V. 50. - P. 2745-2750.

43. Ryabitskii, E. A. Thermocapillary instability of a plane layer with a vertical temperature gradient / E. A. Ryabitskii // Fluid Dynamics. - 1992. - V. 27 - P. 313 –316.

44. Velarde, M. G. Interfacial oscillations in Benard Marangoni layers / M.G. Velarde, P.L. Garcia-Ybarra, J.L. Castillo // Physico Chem. Hydrodyn. – 1987 - V. 9 - P. 387 –392.

45. Hashim I., Wilson SK The onset of oscillatory Marangoni convection in a semi-infinitely deep layer of fluid / I. Hashim, SKWilson // Zeitschrift fur angewandte Mathematik und Physik.– 1999. - V. 50. - P. 546 – 558.

46. Sarychev, V. D. Features of surface doping with pulsed plasma flows of electrically exploded conductors / V. D. Sarychev, V. A. Petrunin, E. A. Budovskikh, P. S. Nosarev,

A.E. Averson // University proceedings. Ferrous metallurgy. - 1991. - No. 4. - S. 64 – 67.

47. Conn Justin, J.A. Fluid dynamical model for antisurfactants / J.A. Conn Justin, R. Duffy Brian, D. Pritchard, S.K. Wilson, P. J. Halling, S. Khellil // Phys. Rev. E - 2016. - V. 93 - P. 043121.

48. Sarychev, V.D. Thermocapillary model of formation of surface nanostructure in metals at electron beam treatment / V.D. Sarychev, S.A. Nevskii, S.V. Konovalov, I.A. Komissarova, E.V. Chermushkina // IOP Conf. Ser. Mater. Sci. Eng. –2015. - V. 91. - P. 012028.

49. Nevskii, S. A. Mathematical model of nanostructure formation in binary alloys at electron beam treatment / S. A. Nevskii, V. D. Sarychev, S. V. Konovalov, D.A. Kosinov, I. A. Komissarova // Materials Science Forum. - 2016. - V.870. - P. 34 – 39.

50. Granovsky, A.Yu. Model for the formation of internal nanolayers in shear flows of materials / A.Yu. Granovsky, V.D. Sarychev, V.E. Gromov // Journal of Technical Physics. - 2013. - V. 83. - Issue. 10 . P.155 – 158.

51. Konovalov, S.V. Mathematical modeling of the concentrated energy flow effect on metallic materials / S.V. Konovalov, Xizhang Chen, V. D. Sarychev, S. A. Nevskii, V. E. Gromov, Milan Trtica // Metals. - 2017– V.7. - P. 1–18.

52. Angles, A.A. Self-oscillatory processes under the influence of concentrated energy flows / A. A. Uglov, S. V. Selishchev. - M .: Nauka, 1987 .

53. Harutyunyan, R.V. The effect of laser radiation on materials / R.V. Harutyunyan, V.Yu. Baranov, L.A. Bolshov et al. - Moscow: Nauka, 1989 .

54. Bunkin, F.V. Nonresonant interaction of high-power optical radiation with a liquid / F.V. Bunkin, M. I. Tribelskii // Sov. Phys. Usp. - 1980. - V. 23. - P. 105 – 133.

55. Samokhin, A.A. Laser vaporization of absorbing liquid under transparent cover / A.A. Samokhin, N.N. Il'ichev, P.A. Pivovarov, A.V. Sidorin // Bulletin of the Lebedev Physics Institute. - 2016 .- V.43 (5). - P. 156 – 159.

56. Akhmanov, S.A. Interaction of powerful laser radiation with the surfaces of semiconductors and metals: nonlinear optical effects and nonlinear optical diagnostics / S.A. Akhmanov, V.I. Emel'yanov, N.I. Koroteev, V.N. Seminogov // Sov. Phys. Usp. 1985 - V. 28. - P. 1084 – 1124.

57. Burelbach, J. P. Nonlinear stability of evaporating / condensing liquid films / J.P. Burelbach, S.G. Bankoff, S.H. Davis // Journal of Fluid Mechanics. - 1988. - V. 195. - P. 463 – 494.

58. Joo, S.W. Long-wave instabilities of heated falling films: two-dimensional theory of uniform layers / S. W. Joo, S. H. Davis // J. Fluid Mech. 1991 .-V. 230 - P. 117 – 146.

59. Samokhin A.A. Influence of evaporation on the melt behavior during laser interaction with metals / A.A. Samokhin // Soviet Journal of Quantum Electronics. - 1983. - V. 13. - P. 1347–1350.

60. Oron, A. Longscale evolution of thin liquid films / A. Oron, S.H. Davis, S.G. Bankoff // Rev. Mod. Phys. –1997. - V. 69. - P. 931 – 980.

61. Mirzade, F. Kh. Wave instability of a molten metal layer formed by intense laser irradiation / F. Kh. Mirzade // Technical Physics. - 2005. - V. 50. - P. 993 – 998.

62. Kuznetsov. V.V. Heat and mass transfer on a liquid–vapor interface / V.V. Kuznetsov // Fluid Dyn. 2011. - V. 46. - P. 754 – 763.

63. Das K.S. Surface thermal capacity and its effects on the boundary conditions at fluid-fluid interfaces / K.S. Das, C.A. Ward // Phys. Rev. E - 2007. - V. 75. - P. 065303–1 - 065303-4.

64. Bekezhanova, V.B. Stability of two-layer fluid flows with evaporation at the interface / V.B. Bekezhanova, O.N. Goncharova, E.B. Rezanova, I.A. Shefer // Fluid Dyn. -

2017. - V. 52. - P. 189 – 200.

65. Iorio, C. S. Student of evaporative convection in an open cavity under shear stress flow / C.S. Iorio, O.N. Goncharova, O.A. Kabov // Microgravity Sci. Technol. - 2009. - V. 21. - P. 313 – 319.

66. Iorio, C.S. Heat and mass transfer control by evaporative thermal pattering of thin liquid layers / C.S. Iorio, O.N. Goncharova, O.A. Kabov // Computational Thermal Sci. - 2011 - V.3. - P. 333 – 342.

67. Goncharova, O.N. Modeling of two-layer fluid flows with evaporation at the interface in the presence of the anomalous thermocapillary effect / O.N. Goncharova, E.V. Rezanova // Journal of Siberian Federal University – Mathematics and Physics. - 2016. - V. 9. - P. 48 – 59.

68. Zuo, J.M. Advanced Transmission Electron Microscopy / J.M. Zuo, J.C. Spence. - New York: Springer, 2017 .-729 p.

69. Egerton, R. Physical Principles of Electron Mcroscopy / R. Egerton. – New York: Springer, 2016 .- 196 p.

70. Sarychev, V.D. Modeling of the initial stages of the formation of heterogeneous plasma flows in the electric explosion of conductors / V.D. Sarychev, S.A. Nevskii, S.V. Konovalov, A. Yu. Granovskii // Current Applied Physics. - 2018. - No. 18. - P. 1101–1107.

Changes in structural phase states and properties of surface layers of Al–Si alloys after electron–ion–plasma effects

5.1 Structure and properties of the Al–Si alloy subjected to complex electron-ion-plasma treatment in various modes

5.1.1 Changes in the structure of the Al–Si alloy after complex processing in mode No. 1

In this section, mode No. 1 of complex electron–ion–plasma processing is considered (Table 2.3), namely:

PMPJ: the mass of the Al–Ti conductor is 58.9 mg, the mass of the Y_2O_3 powder sample is 58.9 mg, the discharge voltage of the capacitor bank is 2.8 kV;

IPEB: accelerated electron energy U = 17 keV, electron beam energy density E_s = 25 J/cm², pulse duration τ = 150 μs, number of pulses n = 3, pulse repetition rate f = 0.3 s⁻¹; residual gas pressure (argon) in the working chamber of the installation p = 2 · 10⁻² Pa.

Typical electron microscopic images of the surface structure of the test material subjected to the combined treatment are shown in Fig 5.1. The presence of a large number of microcraters (Fig 5.1 *a, b*, the microcraters are indicated by arrows) and droplet particles (Fig 5.1 *c*, the particles are indicated by arrows) is clearly visible. The forming surface layer is divided into regions whose sizes do not exceed 1 μm (Fig 5.1 *d*). Regions have a polycrystalline structure; crystallite sizes vary from 60 nm to 100 nm (Fig 5.1 *d*, inset).

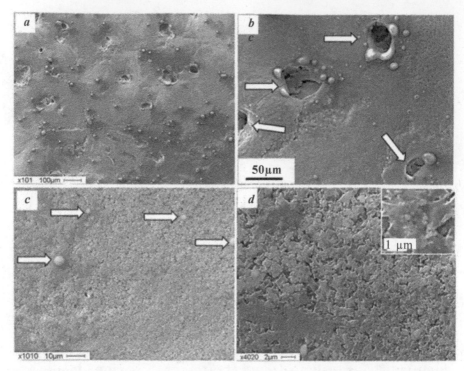

Fig. 5.1. The surface structure of the sample of Al–Si alloy subjected to complex processing. On (*b*) the arrows indicate microcraters; on (*c*) the particles of a droplet fraction.

The elemental composition of the surface layer of an Al–Si alloy modified according to mode No. 1 was studied by X-ray microanalysis. The results of the studies showed that in the surface layer the average concentration of titanium atoms is 17.6 wt. %, yttrium 14.3 wt.%, oxygen 6.7 wt.%. The concentration of yttrium and oxygen atoms in the particles of the droplet fraction is significantly higher (Fig 5.2).

The structure and elemental composition of the surface layer of the modified Al–Si alloy was studied by studying transverse sections. Analyzing the results presented in Fig 5.3, it can be noted that the thickness of the modified layer varies from 50 μm to 70 μm. The modified layer has a submicrocrystalline structure and is free from silicon inclusions and intermetallic compounds present in the volume of the samples under study.

The methods of X-ray microspectral analysis were used to study the distribution of atoms of chemical elements of the Al–Si alloy over the thickness of the modified layer.

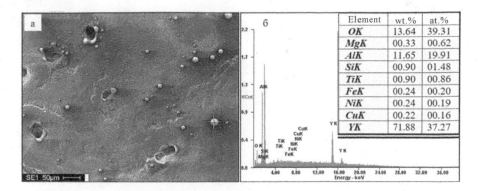

Element	wt.%	at.%
OK	13.64	39.31
MgK	00.33	00.62
AlK	11.65	19.91
SiK	00.90	01.48
TiK	00.90	00.86
FeK	00.24	00.20
NiK	00.24	00.19
CuK	00.22	00.16
YK	71.88	37.27

Fig. 5.2. Structure (*a*) and energy spectra (*b*) obtained by X-ray microanalysis of the microdrops indicated on (*a*) by the (+) sign. Complex modification, mode No. 1

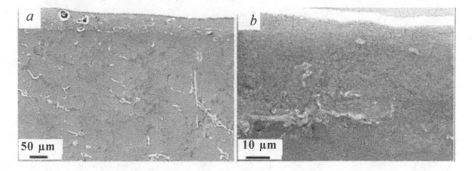

Fig. 5.3. A characteristic electron microscopic image of the structure of a transverse section subjected to complex processing.

The results presented in Fig. 5.4 show that the maximum concentration of titanium, yttrium and oxygen, additionally introduced into the alloy, is concentrated in the surface layer of the sample with a thickness of 70–80 μm. As we move away from the surface of the modification, the concentration of these elements decreases. The modified layer is characterized by a uniform distribution of chemical elements. Outside this layer, there are areas with a high content of individual elements (for example, silicon, nickel, iron and copper) (Fig. 5.4 *a*). It should also be noted that the concentration of titanium in the modified layer is significantly higher than the concentration of yttrium.

Thus, a complex treatment combining PMPJ of an Al–Si alloy with titanium and yttrium oxide and subsequent irradiation with IPEB at an electron beam energy density of 25 J/cm² is accompanied by

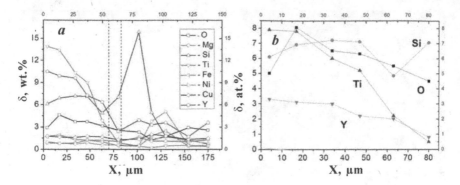

Fig. 5.4. Distribution of the relative content of elements over the thickness of the modified layer.

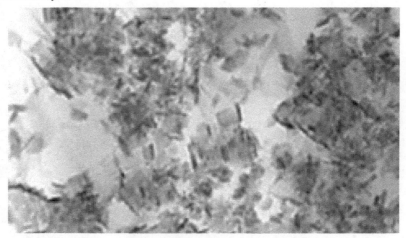

Fig. 5.5. Image of the structure of the surface layer of Al–Si alloy.

the formation of a modified layer up to 70 μm thick enriched with titanium, yttrium and oxygen atoms.

The elemental and phase composition, defective substructure of the Al–Si alloy subjected to combined surface treatment, were studied by transmission electron diffraction microscopy. For this purpose, foils were prepared from plates cut from a massive sample perpendicular to the surface of modification using ion thinning methods, which made it possible to analyze the changes in the elemental composition and the structural phase state of the material depending on the distance from the surface of the modification. A typical image of the structure of the surface layer obtained by the STEM method is shown in Fig. 5.5. It is clearly seen that the modified layer is formed by crystallites of various morphologies, the sizes of which vary in the submicron–nanometer range.

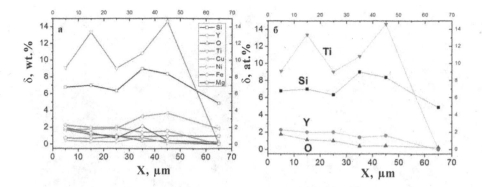

Fig. 5.6. Dependence on the distance to the surface of the modification of the concentration of chemical elements (excluding aluminium) of samples subjected to complex processing; transmission electron microscopy (STEM analysis method).

The results of the study of the elemental composition of the surface layer of the studied material subjected to combined treatment are shown in Fig. 5.6. The dimensions of the studied volume of the foil are $9.5 \times 9.5 \times 0.3$ μm³.

By analyzing the results presented in Fig. 5.6, it can be noted that the thickness of the layer doped with yttrium, oxygen, and titanium is (60–70) μm. At a greater distance from the surface of the modification, the concentration of these elements was negligible. The elemental composition of the modified layer depends on the distance to the surface of the modification. Most significantly, as the distance from the treatment surface decreases, the concentration of oxygen and yttrium atoms decreases.

The main alloying element of the modified layer is titanium, whose average concentration in the layer is ≈11 at. % and ranges from (9–14.5) at. %, showing a tendency to increase with distance from the surface of the alloying. The relative content of yttrium and oxygen decreases monotonically with increasing distance from the doping surface.

The distribution of alloying elements in the modified layer was studied by mapping methods [1]. The results of mapping a surface layer with a thickness of 10 μm are shown in Fig. 5.7. The results of a quantitative analysis of the elemental composition of this layer are presented in Table 5.1 (layer (0–10) μm).

The volume of inclusions is enriched in titanium atoms, yttrium and silicon atoms form the shell of these inclusions. In some cases, copper atoms are present in the shell. A similar structure is detected in a layer with a thickness of ≈60 μm. At a greater distance from

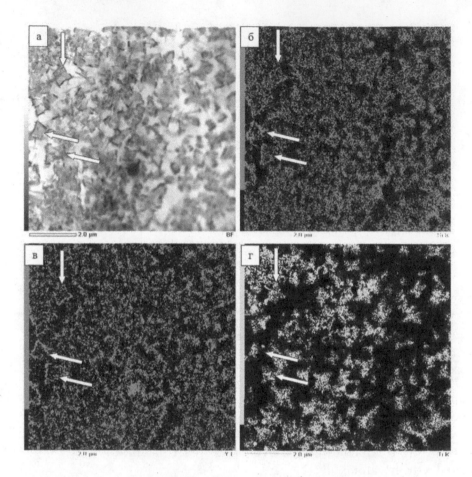

Fig. 5.7. Electron microscopic image (*a* – bright field) of the surface layer of Al–Si alloy subjected to complex processing; b–d – images of this foil layer obtained in the characteristic x-ray of silicon atoms (*b*), yttrium (*c*) and titanium (*d*)

Table 5.1. The chemical composition (at.%) of the alloy layers located at different distances (*X*) from the surface of the complex treatment

X, μm	Al	Si	Y	O	Ti	Cu	Ni	μ
0–10	76.03	6.77	2.26	1.73	9.1	1.72	0.45	1.94
10–20	73.0	7.01	1.99	1.15	13.37	1.79	0.32	1.36
20–30	78.71	6.35	2.01	1.0	9.03	1.88	0.29	0.72
30–40	72.03	9.0	1.42	0.42	10.84	3.34	0.8	2.15
40–50	70.73	8.36	1.59	0.43	14.6	3.7	0.16	0.42
60–70	92.87	4.88	0.0	0.26	0.06	1.85	0.04	0.04

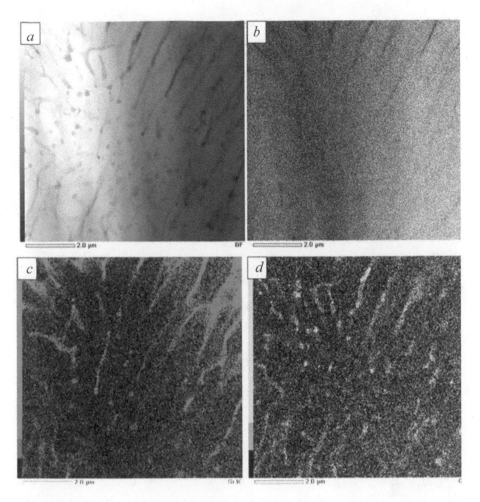

Fig. 5.8. Electron microscopic image (*a* – bright field) of a layer located at a distance of (60–70) μm from the surface of the complex treatment; *b, d* – images of this foil layer obtained in the characteristic X-ray radiation of aluminium atoms (*b*), silicon (*c*) and copper (*d*).

the surface of the modification, the structure of high-speed cellular crystallization is observed, the main alloying elements of which are silicon and copper (Fig. 5.8).

The cell volume is formed by an aluminium-based solid solution (Fig. 5.8 *b*), along the cell boundaries there are extended interlayers enriched with silicon and copper atoms (Fig. 5.8 *c, d*). The thickness of such layers varies from 50 nm to 250 nm. At a greater distance from the surface of the modification, structures characteristic of the cast state of the material are revealed, i.e. solid solution grains based on aluminium, eutectic, silicon inclusions and intermetallic compounds of various elemental composition are observed.

Using dark-field images of transmission electron diffraction microscopy and the method of decoding microelectron diffraction patterns [2–6], the phase composition of the modified layer is analyzed. Figure 5.9 *a* shows an electron microscopic bright-field image of this layer.

The microelectron diffraction pattern obtained from the portion of the foil highlighted by the selector diaphragm (Fig. 5.9 *b*) contains a large number of reflexes of various intensities (Fig. 5.9 *c*). The decoding of the microelectron diffraction pattern revealed reflections of the following phases: silicon, α-titanium, SiY, SiTi, Cu_2YSi_2. Reflections belonging to the silicon crystal lattice form diffraction rings (Fig. 5.9 *c*, reflex 1), which indicates the small particle sizes of this phase. Indeed, the dark-field image obtained in the reflections of the [111] Si ring demonstrates the presence of nanoscale (10–20 nm) particles in the structure of the material (Fig. 5.9 *d*). The most intense reflex of microelectron diffraction patterns (Fig. 5.9 *c*, reflex 2) corresponds to [101] α-Ti. The dark-field image (Fig. 5.9 *e*) of the foil obtained in this reflex indicates that the faceted particles are formed by α-titanium. The most difficult to interpret is the dark-field image obtained in closely spaced reflexes indicated in Fig. 5.9 *c*, number 3.

Analysis of the microelectron diffraction pattern suggests that these reflections belong to the phases SiY, Cu_2YSi_2, and SiTi, which, judging by the dark-field image shown in Fig. 5.9 *f*, form a shell of particles of α-titanium.

5.1.2. Structural transformations in the surface layers of the Al–Si alloy after complex processing in mode No. 2

In this section, mode No. 2 of complex, electron-ion-plasma processing (Table 2.3) is considered:

PMPJ: the mass of the Al–Ti conductor is 58.9 mg, the mass of the Y_2O_3 powder sample 88.3 mg, the discharge voltage of the capacitor bank is 2.6 kV;

IPEB: accelerated electron energy $U = 17$ keV, electron beam energy density $E_S = 25$ J/cm^2, pulse duration $\tau = 150$ μs, number of pulses $n = 3$, pulse repetition rate $f = 0.3$ s^{-1}; residual gas pressure (argon) in the working chamber of the installation $p = 2 \cdot 10^{-2}$ Pa.

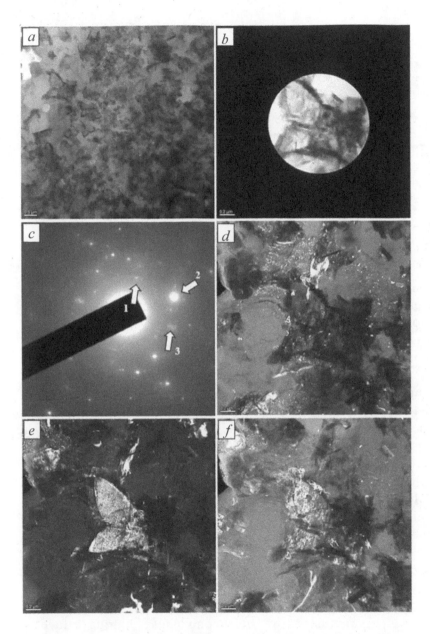

Fig. 5.9. Electron-microscopic image of the structure of the surface layer of the alloy subjected to complex processing; *a, b* – bright fields; *c* – microelectron diffraction pattern (arrows indicate reflections in which dark fields were obtained: 1 – (*d*), 2 – (*e*), 3 – (*f*)); *d–f* are the dark fields obtained in the reflections of [111] Si (*d*), [101] α-Ti (*e*), [040] SiY + [103] Cu_2YSi_2 + [210] SiTi (*f*), respectively. Figure (*b*) shows the foil region highlighted by the selector diaphragm from which the microelectron diffraction pattern was obtained.

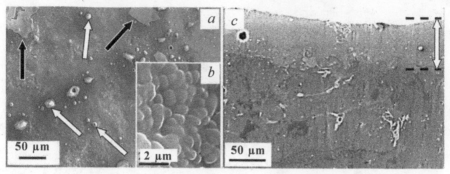

Fig. 5.10. The structure of the sample subjected to PMPJ and subsequent IPEB; *a, b* — surface structure of irradiation; *c* – the structure of the transverse etched thin section. On (*a*) the dark arrows indicate the films, the light arrows indicate microdrops located on the surface of the sample; on (*c*) - a modified layer.

Typical electron microscopic images of the surface structure of the samples subjected to complex processing according to this mode are shown in Fig. 5.10.

It is clearly seen that as a result of processing, a surface is formed containing microcraters, microdrops and the formation of a film form (Fig. 5.10 *a*). The resulting surface layer has a submicrocrystalline structure; crystallite sizes do not exceed 1 μm (Fig. 5.10 *b*). An analysis of the structure of the etched transverse sections showed that the thickness of the modified, as a result of complex processing, layer is (70–80) μm (Fig. 5.10 *c*).

The elemental composition, phase morphology, and the state of the defective substructure of the Al–Si alloy at different distances from the processing surface were studied by transmission electron diffraction microscopy.

The analysis presented in Fig. 5.11 images shows that a cellular crystallization structure is formed in a layer up to 80 μm thick. The cell size varies from 0.8 μm to 1.3 μm. The cells are separated by interlayers of the second phase. The thickness of the interlayers varies within (50–75) nm. Mainly in the triple joints of the cell boundaries are inclusions of the second phase, having a faceted shape in the form of a cuboid or four-petalled rosettes. The sizes of such inclusions vary within (0.5–0.7) μm. Thus, complex processing according to the 2nd regime leads to the formation of a surface layer, the inclusions of the second phase of which are many times (tens to hundreds of times) fewer than inclusions present in the material in the cast state.

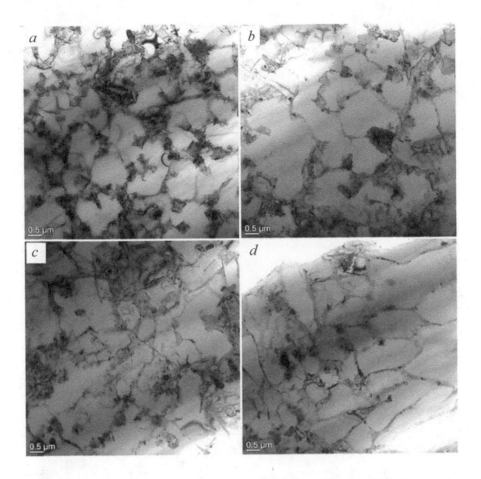

Fig. 5.11. The structure of the alloy subjected to complex modification: a–d – layers located at a distance of 20 µm; 40 µm; 65 µm; 80 µm from the processing surface, respectively

The distribution of chemical elements in the modified layer was studied by X-ray microanalysis of thin foils. The results of elemental analysis (mapping method) of a layer adjacent to the modified surface are shown in Fig. 5.12. It is clearly seen that the interlayers located at the boundaries of the cells of high-speed crystallization are enriched with silicon and yttrium atoms. The faceted particles are enriched in titanium atoms. Yttrium atoms form thin films and droplets located on the surface of the sample modification.

In quantitative terms, the elemental composition of the surface layer, the image of which is shown in Fig. 5.12 a, is given in Table 5.2.

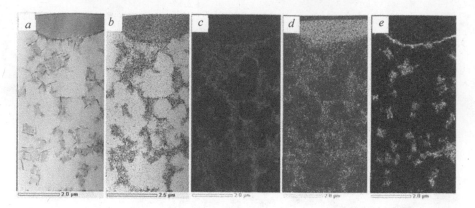

Fig. 5.12. Bright-field electron microscopic image of the structure of the surface layer of the modified Al–Si alloy (*a*); *c–d* — images of this layer obtained in the characteristic x-ray emission of silicon atoms (*c*), yttrium (*d*) and titanium (*e*); image (*b*) obtained by overlaying images (*c–e*).

Table 5.2. The chemical composition of the surface layer of the material subjected to complex processing according to the second mode

Element	(keV)	Mass %	At. %
O	0.525	2.30	4.17
Mg	1.253	1.13	1.35
Al	1.486	75.76	81.60
Si	1.739	5.01	5.18
Ti	4.508	5.14	3.12
Fe	6.398	2.49	1.30
Ni	7.471	1.07	0.53
Cu	8.040	3.33	1.52
Y	1.922	3.78	1.23
Total		100.00	100.00

Analyzing the results of this table, it can be noted that the main chemical element of the studied layer is aluminium, the mass fraction of which exceeds 75%. The concentration of the identified alloying elements varies from 1 wt.% to 5 wt.%

The elemental composition of the modified layer revealed by electron microscopy of thin foils depends on the distance to the processing surface, which follows from the analysis of the results presented in Fig. 5.13. In the most significant way, as the distance

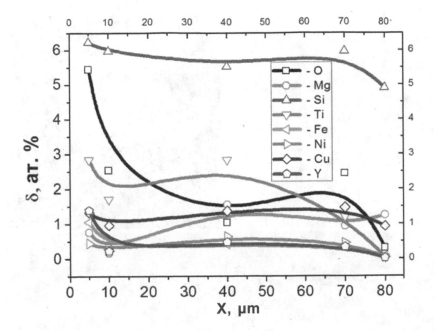

Fig. 5.13. Dependence of the concentration of alloying elements on the distance to the modification surface.

from the treatment surface decreases, the concentration of oxygen and titanium atoms decreases.

The phase composition of the modified layer was analyzed using dark-field images and the method of decoding microelectron diffraction patterns [2–6] by transmission electron diffraction microscopy, Figure 5.14 *a* shows an electron microscopic bright-field image of the surface layer of a modified Al–Si alloy. The microelectron diffraction pattern obtained from the portion of the foil highlighted by the selector diaphragm (Fig. 5.14 *b*) contains a diffraction halo corresponding to the amorphous state of the substance and reflexes forming diffraction rings (Fig. 5.14 *c*). An analysis of the microelectron diffraction pattern revealed the reflections of silicon and yttrium silicide of the SiY composition. Following the results of X-ray microanalysis of this foil section, presented in Fig. 5.14 *d*, it can be assumed that the amorphous phase is the portion of the surface of the sample (film or drop) enriched in yttrium. One of the phases with a nanocrystalline structure and located along the

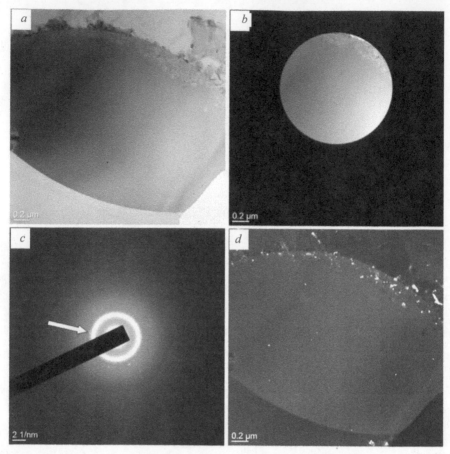

Fig. 5.14. Electron microscopic image of the structure of the surface layer of the sample subjected to complex processing according to the 2nd embodiment; *a, b* bright fields; *c* – microelectron diffraction pattern (the arrow indicates the reflex in which the dark field is obtained); *d* is the dark field obtained in the [211] Si reflection. Figure (*b*) shows the foil region highlighted by the selector diaphragm from which the microelectron diffraction pattern was obtained (*c*).

interface between the droplet and the bulk of the sample is yttrium silicide with the composition SiY.

An electron microscopic image of the cellular crystallization structure of the modified layer is shown in Fig. 5.15. It is clearly seen that the volume of the cells of high-speed crystallization is formed by a solid solution based on the aluminium crystal lattice. Interlayers separating crystallization cells contain silicon particles.

An electron microscopic image of particles in the form of four-petal sockets is shown in Fig. 5.16. Using dark-field analysis methods, it was shown that these particles are α-titanium.

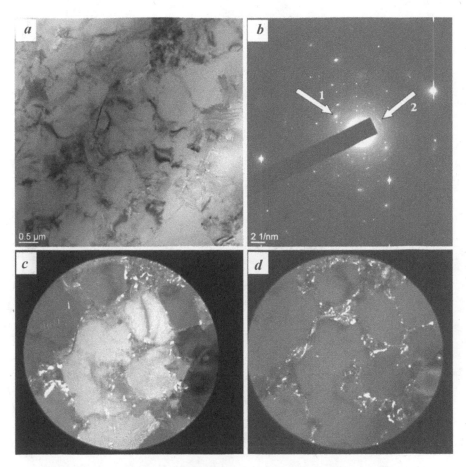

Fig. 5.15. Electron microscopic image of the structure of cellular crystallization of the surface layer of the sample subjected to complex processing according to the 2nd mode; *a* – bright field; *b* – microelectron diffraction pattern (arrows indicate reflexes in which dark fields are obtained: 1 – (*c*), 2 – (*d*)); *c, d* – dark fields obtained in the reflections of [111] Al (*c*) and [111] Si (*d*), respectively.

Thus, as a result of the performed studies, it was shown that as a result of complex processing of the Al–Si alloy, a multi-element multiphase layer with a thickness of ≈80 μm having a submicro-nanocrystalline structure is formed. The presence of droplets enriched in yttrium atoms in an amorphous state was detected on the surface of the modification. It was shown that high-speed crystallization of the doped surface layer is accompanied by the formation of α-titanium particles in the form of cuboids and four-petalled rosettes.

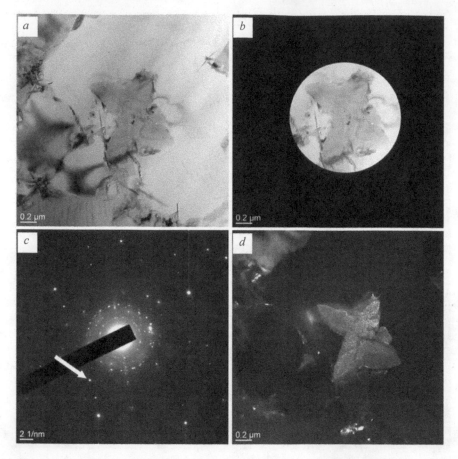

Fig. 5.16. Electron microscopic image of the structure of the surface layer of the sample subjected to complex processing according to the 2nd mode; *a, b* – bright fields; *c* – microelectron diffraction pattern (the arrow indicates the reflex in which the dark field is obtained); *d* – dark field obtained in the reflection [104] α-Ti (*d*).

5.1.3. Evolution of the structural phase states of an Al–Si alloy after complex processing in mode No. 3

In this section, mode No. 3 of complex electron-ion-plasma processing (Table 2.3) is considered:

PMPJ: the mass of the Al–Ti conductor 58.9 mg, the mass of the Y_2O_3 powder sample 58.9 mg, the discharge voltage of the capacitor bank 2.8 kV;

IPEB: accelerated electron energy $U = 17$ keV, electron beam energy density $E_s = 35$ J/cm^2, pulse duration $\tau = 150$ μs, number of pulses $n = 3$, pulse repetition rate $f = 0.3$ s^{-1}; residual gas pressure (argon) in the working chamber of the installation $p = 2 \cdot 10^{-2}$ Pa.

A typical electron microscopic image of the surface structure of a sample subjected to combined treatment in this mode is shown in Fig. 5.17. It is clearly seen that as a result of complex processing, a relief surface is formed containing areas that differ in contrast (Fig. 5.17 *a*). The latter may indicate the heterogeneity of the elemental composition of the surface layer of the material. The formed surface layer has a submicrocrystalline structure, the crystallite sizes of which do not exceed 1 µm (Fig. 5.17 *b*).

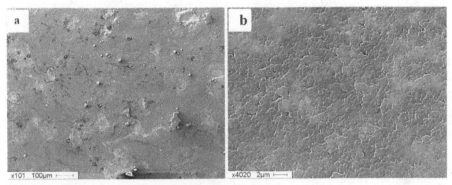

Fig. 5.17. Structure of the modified surface of Al–Si alloy.

The results of studies performed by the methods of X-ray microspectral analysis of the studies showed that the average concentration of yttrium in the surface layer of the investigated material is 8.3 wt.% (Fig. 5.18).

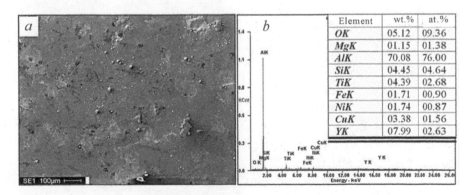

Element	wt.%	at.%
OK	05.12	09.36
MgK	01.15	01.38
AlK	70.08	76.00
SiK	04.45	04.64
TiK	04.39	02.68
FeK	01.71	00.90
NiK	01.74	00.87
CuK	03.38	01.56
YK	07.99	02.63

Fig. 5.18. Structure (*a*) and energy spectra (*b*) obtained by X-ray microanalysis of the sample site (*a*),

186

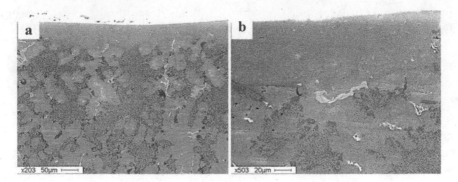

Fig. 5.19. A characteristic electron microscopic image of the structure of a transverse section subjected to a combined treatment in the third mode.

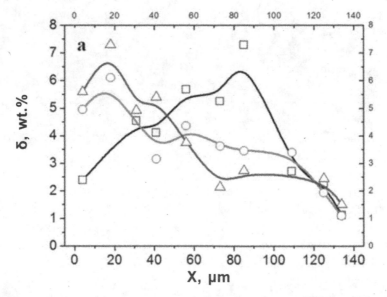

Fig. 5.20. Distribution of the relative content of yttrium atoms over the thickness of the modified layer. The results were obtained by averaging the data revealed over three tracks of elemental analysis.

Analyzing the results, cross-sectional studies presented in Fig. 5.19, it can be noted that the thickness of the modified layer varies from 45 μm to 80 μm. The modified layer has a submicron nanocrystalline structure and is free from inclusions of silicon and intermetallic compounds present in the material under study.

The results of the study, the distribution of yttrium atoms over the thickness of the modified layer presented in Fig. 5.20, show that when the combined treatment is performed, the concentration profile

Fig. 5.21. Electron microscopic image of the structure of the alloy subjected to complex electron–ion–plasma treatment according to the third mode; *a* – structure of the surface layer; *b* – the structure of the layer located at a depth of (20–30) μm.

has one maximum, the position of which depends on the analyzed area of the sample.

Analysis of the concentration profiles shown in Fig. 5.20, indicates that the heterogeneity of the distribution of yttrium atoms is detected both in the transverse and longitudinal sections of the material, i.e. is voluminous in nature.

Using transmission electron diffraction microscopy methods, it was possible to detect the formation in the modified layer of a gradient submicron-sized structure, a typical image of which is shown in Fig. 5.21.

It was found that the modified layer with a thickness of up to 70 μm has a high-speed cellular crystallization structure. Cell sizes vary from 0.5 μm to 1.2 μm. The cells are separated by interlayers of the second phase (Fig. 5.21 *b*). The structure of the surface layer contains faceted inclusions (Fig. 5.21 *a*, dark coloured inclusions), the sizes of which vary from 0.4 μm to 0.8 μm. The relative content of such inclusions decreases with distance from the surface of the modification.

The energy spectra obtained by X-ray microanalysis of thin foils from a surface modified layer are shown in Fig. 5.22. The results of the quantitative analysis of elemental composition are presented in Table 5.3.

Analyzing the results presented in Table 5.3, it can be noted that the surface layer is multi-element and, along with the atoms of the

Table 5.3. X-ray microanalysis of the elemental composition of the surface layer of an Al–Si alloy

Element	(keV)	wt.%	at.%
O	0.525	0.58	1.04
Al	1.486	79.97	85.77
Si	1.739	7.29	7.52
Ti	4.508	3.77	2.28
Cr	5.411	0.11	0.06
Fe	6.398	0.61	0.32
Ni	7.471	0.75	0.37
Cu	8.040	3.41	1.55
Y	1.922	2.64	0.86
Ag	2.984	0.88	0.24
Total		100.00	100.00

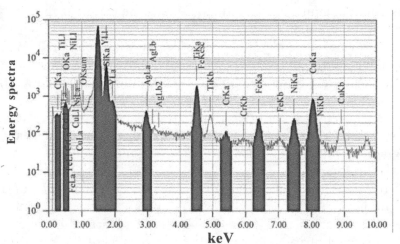

Fig. 5.22. Energy spectra obtained by X-ray microanalysis of the surface layer of Al–Si alloy.

starting material (aluminium, silicon, copper, nickel, chromium, iron), is additionally enriched with atoms of titanium, yttrium and oxygen.

The mapping method allows an analysis of the distribution of alloying elements in the investigated volume of material. The results of mapping the surface layer of the modified alloy are shown in Fig. 5.23.

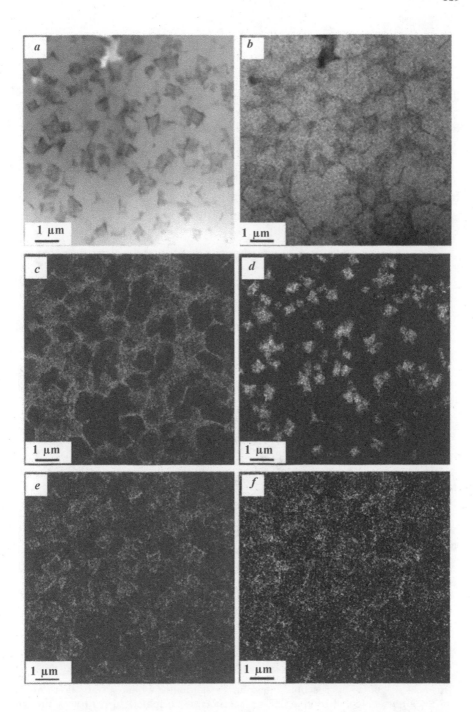

Fig. 5.23. Electron microscopic image of the structure of the alloyed layer (*a*); images (*b–f*) were obtained in the characteristic X-ray radiation of atoms of aluminium (*b*), silicon (*c*), titanium (*d*), yttrium (*e*) and copper (*f*).

It can be seen that the cells of high-speed crystallization are enriched mainly by aluminium atoms (Fig. 5.23 *b*). The cells are separated by layers enriched mainly by silicon atoms (Fig. 5.23 *c*). Faceted inclusions (Fig. 5.23 *a*, dark coloured inclusions) are enriched mainly in titanium, aluminium and copper atoms (Fig. 5.23 *c*, *d*, *f*), yttrium atoms mainly form interlayers along the boundaries of the faceted inclusions (Fig. 5.23*d*) .

Figure 5.24 shows the results of the analysis of the phase composition of the surface layer of the foil section containing faceted inclusions. We used a technique based on obtaining dark-field images and a technique for displaying microelectron diffraction patterns [2–6].

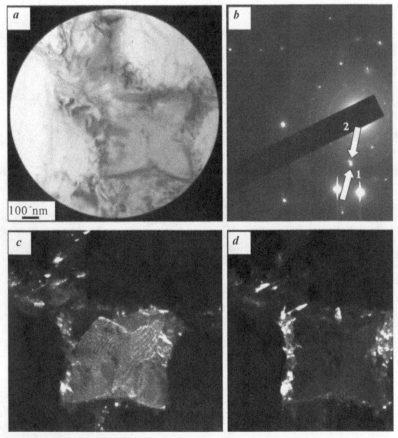

Fig. 5.24. Electron-microscopic image of the structure of the surface layer of an Al–Si alloy subjected to complex processing; *a* – bright field (foil area limited by the selector diaphragm); *b* – microelectronogram corresponding to the bright field; *c*, *d* – dark fields obtained in reflections of [200] Al$_5$CuTi$_2$ and [300] AlCuY, respectively. On (*b*) arrows indicate reflexes in which dark fields are obtained: 1 – (*c*); 2 – (*d*).

Electron microscopic microdiffraction analysis shows that the faceted inclusions are formed by the phase of the Al_5CuTi_2 composition (Fig. 5.24 *c*). Along the boundaries of these inclusions, interlayers with AlCuY phase composition are revealed (Fig. 5.24 *d*).

Figure 5.25 shows a characteristic image of the structure of cellular crystallization. The microelectron diffraction pattern obtained from this portion of the foil contains separately located point reflexes and reflexes forming rings (Fig. 5.25 *c*). Indication of the microelectron diffraction pattern showed that the reflections

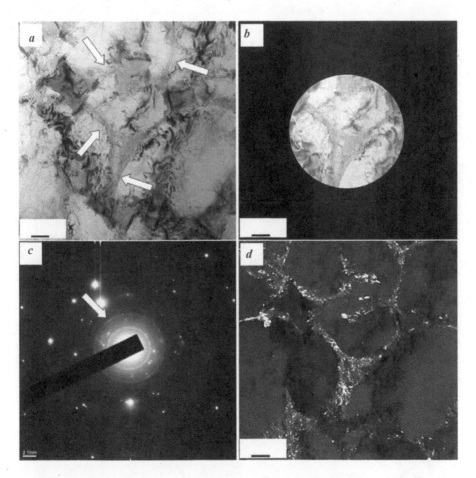

Fig. 5.25. Electron-microscopic image of the structure of the surface layer of Al–Si alloy subjected to complex processing in the third mode; *a, b* – bright fields; *c* – microelectron diffraction pattern obtained from the foil portion limited by the selector diaphragm (the portion image is shown in (*b*)); *d* is the dark field obtained in the [220] Si reflex indicated on (*c*) by the arrow. Arrows on (*a*) indicate silicon interlayers.

forming the diffraction rings belong to the silicon lattice. A dark-field image of the structure of the surface layer of the studied material obtained in reflections of the diffraction ring (Fig. 5.25 c, reflexes are indicated by an arrow) is shown in Fig. 5.25 d. By analyzing the results presented in Fig. 5.25 d, it can be noted that silicon interlayers located along the boundaries and at the joints of the boundaries of crystallization cells formed by an aluminium-based solid solution have a nanocrystalline structure with crystallite size varying within (10–20) nm.

Thus, the complex surface treatment of the Al – Si alloy in regime 3 led to a radical transformation of the structure of the surface layer of the material with a thickness of ≈70 μm, consisting in the dissolution of silicon inclusions and intermetallic compounds characteristic of the cast state of the material under study, and the formation of a gradient multielement submicro-nanoscale structure. It was found that the modified layer has a high-speed cellular crystallization structure and contains faceted inclusions, the relative content of which decreases with distance from the surface of the modification. It was shown by X-ray microspectral analysis that the surface layer of the alloy is multi-element and, along with the atoms of the starting material (aluminium, silicon, copper, nickel, chromium, iron), is additionally enriched with atoms of titanium, yttrium and oxygen. It was found that cells of high-speed crystallization are enriched mainly by aluminium atoms; interlayers separating cells are enriched mainly in silicon atoms; faceted inclusions are enriched mainly by atoms of titanium, aluminium and copper; yttrium atoms, mainly, form interlayers along the boundaries of faceted inclusions. It was revealed that silicon interlayers located along the boundaries and at the joints of the boundaries of crystallization cells formed by an aluminium-based solid solution have a nanocrystalline structure with crystallite size varying within (10–20) nm.

5.1.4. Analysis of the structure and phase composition of the surface layers of the Al–Si alloy after complex processing in mode No. 4

In this section, mode No. 4 of complex electron-ion-plasma processing (Table 2.3) is considered:

PMPJ: the mass of the Al–Ti conductor 58.9 mg, the mass of the Y_2O_3 powder sample 88.3 mg, the discharge voltage of the capacitor bank is 2.6 kV;

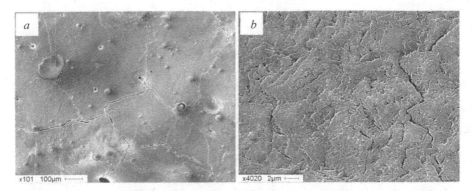

Fig. 5.26. Modified surface structure.

IPEB: accelerated electron energy U = 17 keV, electron beam energy density E_S = 35 J/cm^2, pulse duration τ = 150 μs, number of pulses n = 3, pulse repetition rate f = 0.3 s^{-1}; residual gas pressure (argon) in the working chamber of the installation p = 2 · 10^{-2} Pa.

A typical electron microscopic image of the surface structure of an Al–Si alloy subjected to a combined treatment according to the fourth mode is shown in Fig. 5.26. Fragmentation of the surface layer by microcracks is observed; on the modified surface, micropores, microcraters, and influxes of material are found (Fig. 5.26 a). The microstructure of the surface layer is formed by crystallites with sizes (0.4–0.7) μm (Fig. 5.26 b).

The results of the studies by X-ray microspectral analysis showed that in the surface layer of the material the average concentration of yttrium is 17.9 wt.%, titanium 22.5 wt.%, oxygen 6.3 wt.%.

Analyzing the results of the study of transverse sections presented in Fig. 5.27 a, it can be noted that the thickness of the modified layer

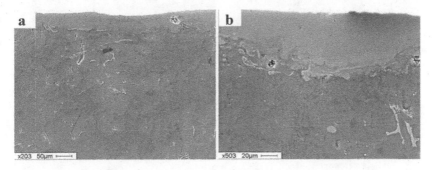

Fig. 5.27. Typical electron microscopic image of the cross-sectional structure after combined processing.

varies from 45 μm to 80 μm. The modified layer has a submicro nanocrystalline structure and is free from inclusions of silicon and intermetallic compounds present in the cast state of the material (Fig. 5.27 *b*).

X-ray spectral analysis methods have been used to study the distribution of yttrium atoms over the thickness of the modified layer. The results presented in Fig. 5.28 show that two maxima of the distribution of yttrium atoms in the volume of the modified layer are fixed. Often the second maximum corresponds to the 'modified layer/bulk material' interface.

The concentration of yttrium depends on the analyzed area of the material. Analysis of the concentration profiles shown in Fig. 5.28, indicates that the heterogeneity of the distribution of yttrium atoms is detected both in the transverse and longitudinal sections of the material, i.e. is voluminous in nature.

The results of X-ray microspectral analysis of the elemental composition of the foil prepared from the Al–Si alloy modified by the complex method are shown in Fig. 5.29. It is clearly seen that the thickness of the doped layer, i.e. the layer in which the presence of alloying elements (titanium, yttrium, oxygen) is detected does not exceed 170 μm. The main elements of this layer are aluminium and titanium. The concentration of the remaining elements varies from 1 wt.% to 5 wt.% With increasing distance from the surface of the modification, the relative content of titanium and yttrium decreases,

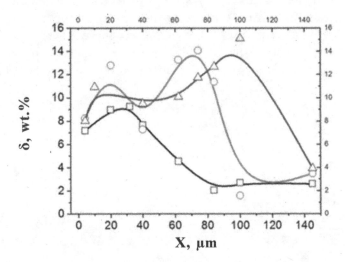

Fig. 5.28. Distribution of the relative content of yttrium atoms over the thickness of the modified layer. The results are obtained for three tracks of elemental analysis.

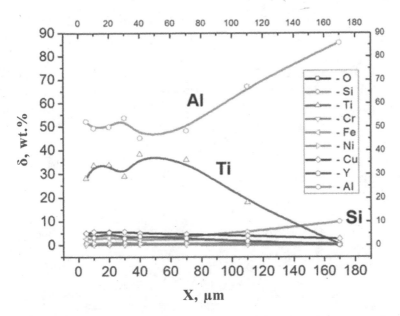

Fig. 5.29. Dependence of the relative content of chemical elements on the distance from the modification surface

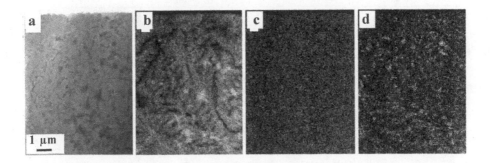

Fig. 5.30. Electron microscopic (bright field) image of the structure of the modified layer adjacent to the processing surface (*a*, the upper part of the image corresponds to the surface of the modification); *b–d* – images of a given portion of the foil obtained in the characteristic x-ray radiation of atoms of titanium (*b*), silicon (*c*) and yttrium (*d*).

the concentration of aluminium and silicon atoms increases, reaching values characteristic of the chemical composition of the cast alloy, the concentration of the remaining alloying elements of the alloy varies slightly.

To visualize the distribution of chemical elements of the modified sample layer allows X-ray spectral analysis, namely, the mapping method.

Figure 5.30 shows the results of a study of the distribution of titanium, silicon, and yttrium atoms in a layer ≈10 μm thick adjacent to the complex treatment surface. It is seen that the atoms of these elements are not uniformly distributed in the surface layer, forming inclusions of various shapes and sizes. It should be noted that a structure of this type is observed in a layer up to 40 μm thick.

The results of a quantitative analysis of the elemental composition of the foil plot shown in Fig. 5.30 *a* are presented in Table 5.4.

Table 5.4. X-ray microanalysis of the foil area shown in Fig 5.30 *a*

Element	(keV)	wt.%	at.%
O	0.525	0.83	1.76
Al	1.486	52.18	65.51
Si	1.739	4.87	5.87
Ti	4.508	28.13	19.89
Cr	5.411	0.26	0.17
Fe	6.398	2.76	1.67
Ni	7.471	0.94	0.54
Cu	8.040	5.04	2.68
Y	1.922	4.99	1.90
Total		100.00	100.00

At a distance of (40–50) μm from the surface of the complex treatment, there is a layer of material containing spherical particles enriched in yttrium and oxygen atoms (Fig. 5.31, the particle is indicated by arrows). The shape of the particles and their elemental composition suggest that these particles are particles of the initial yttrium oxide powder that did not collapse during PMPJ. Particle sizes range from 50 nm to 1.2 μm. The results of X-ray spectral quantitative analysis of this foil section are shown in Table 5.5.

Figure 5.32 shows the results of microdiffraction electron microscopy analysis of the surface layer of an Al–Si alloy (the surface of the modification is indicated by the arrow in Fig. 5.32 *a*).

It can be seen that the sizes of crystallites forming the studied material layer vary from units to hundreds of nanometers, i.e. the modifiable layer is a submicro nanocrystalline material. X-ray

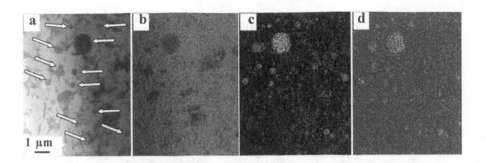

Fig. 5.31. Electron-microscopic bright-field image of the structure of a layer located at a distance of (40–50) μm from the surface of the complex treatment according to the fourth mode (*a*); *b–d* – images of this section of the foil obtained in the characteristic x-ray radiation of aluminium atoms (*b*), yttrium (*c*) and oxygen (*d*). The arrows on (*a*) indicate particles of yttrium oxide.

Table 5.5. X-ray microanalysis of the foil plot shown in Fig 5.31 *a*

Element	(keV)	wt.%	at.%
O	0.525	0.87	1.91
Al	1.486	45.17	59.15
Si	1.739	3.34	4.20
Ti	4.508	38.24	28.20
Cr	5.411	0.29	0.20
Fe	6.398	2.36	1.50
Ni	7.471	0.82	0.49
Cu	8.040	5.10	2.83
Y	1.922	3.82	1.52
Total		100.00	100.00

microanalysis of the surface layer of the modified material showed that the main chemical elements of this layer are aluminium and titanium, silicon, copper and yttrium are present in much smaller amounts (Fig. 5.30, Table 5.4). The results of a dark-field analysis of the phase composition of this layer are shown in Fig. 5.32 *c–f*. Analysis of microelectron diffraction patterns shows that crystallites of submicron sizes are formed by a solid solution based on aluminium (Fig. 5.32 *e*). Inclusions of the nanoscale range are formed by particles of titanium and yttrium aluminides of the composition Al_3Ti and Y_3Al_2, as well as titanium silicides of the composition $TiSi_2$.

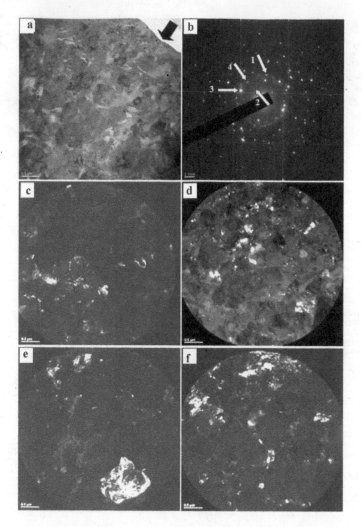

Fig. 5.32. Electron microscopic image of the structure of the layer adjacent to the surface of the modification; a – bright field; b – microelectron diffraction pattern; c – dark fields obtained in reflections of [004] TiSi$_2$, [002] Y$_3$Al$_2$, [111] Al, [118] Al$_3$Ti, respectively; the arrows indicate: on (a) the surface of the modification, on (b) the reflections in which dark fields were obtained (1) – c, (2) – d, (3) – e, 4 – f.

The results of microdiffraction electron microscopic analysis of a layer located at a distance of ≈70 μm are shown in Fig. 5.33.

It is clearly seen that at a given distance from the surface of the modification, the alloy structure is represented by high-speed crystallization cells. Cell sizes vary within (0.5–0.6) μm. Analysis of the microelectron diffraction pattern (Fig. 5.33 c) shows that the crystallization cells are formed by a solid solution based on

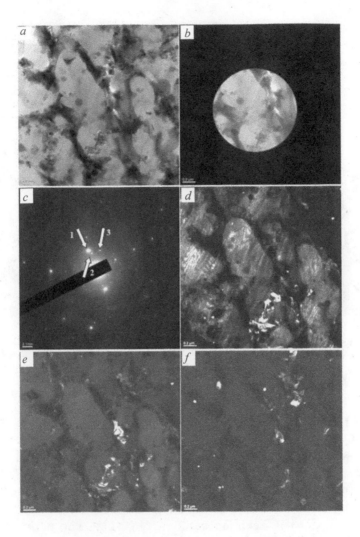

Fig. 5.33. Electron microscopic image of the structure of a layer located at a distance of ≈70 μm from the surface of the modification; a, b – bright fields; c – microelectron diffraction pattern; $d–f$ – dark fields obtained in the reflections of [111] Al + [302] Si, [111] Si, [111] $Cu_{2.7}Fe_{6.3}Si$, respectively; on (c) the arrows indicate the reflexes in which the dark fields are obtained (1) - d, (2) – e, (3) – f.

aluminium (Fig. 5.33 d). The cells are separated by interlayers of the second phase, the transverse dimensions of which vary within (50–70) nm. Microdiffraction analysis using the dark-field imaging technique shows that silicon particles (Fig. 5.33 d, f) and particles of the $Cu_{2.7}Fe_{6.3}Si$ compound (Fig. 5.33 f) are located at the boundaries of the crystallization cells.

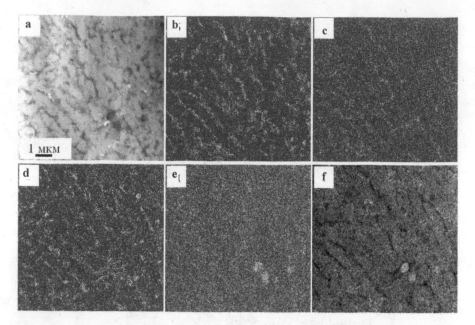

Fig. 5.34. Electron-microscopic bright-field image of the structure of a layer located at a distance of (≈70) μm from the surface of the complex treatment (*a*); *b–f* – images of a given portion of the foil obtained in the characteristic x-ray radiation of atoms of iron (*b*), copper (*c*), yttrium (*d*), silicon (*e*) and titanium (*e*).

The results of X-ray microspectral analysis presented in Fig. 5.34, confirm the possibility of the formation of interlayers of this elemental composition.

Thus, studies performed by scanning and transmission electron diffraction microscopy show that the thickness of the doped layer, i.e. the layer in which the presence of alloying elements (titanium, yttrium, oxygen) is detected reaches ≈170 μm. The main elements of the alloyed layer are aluminium and titanium. Using the mapping method, an inhomogeneous distribution of atoms of alloying elements in the modified layer is revealed. It is shown that PMPJ is accompanied by both doping of the surface layer with plasma elements and the incorporation of particles of the initial yttrium oxide powder into the surface layer. It has been established that complex processing leads to the formation of a multiphase submicron-sized state in the surface layer of Al–Si alloy, the crystallite sizes of which vary from a few to hundreds of nanometers.

5.2. Effect of electron–ion–plasma treatment on the mechanical properties of Al–Si alloys

5.2.1. Change in the microhardness of Al–Si alloy depending on the method and mode of modification

The revealed transformations of the Al–Si alloy surface layer should have a significant effect on the mechanical and tribological properties of the material. As the mechanical properties, the change in microhardness was evaluated depending on the method and modification mode.

The change in microhardness directly on the surface of the material modification, after PMPJ with optimal parameters (mass of powder charge Y_2O_3 58.9 mg, discharge voltage U 2.8 kV and Y_2O_3 88.3 mg, $U = 2.6$ kV) and after IPEB in the optimal modes (25 J/cm^2 and 35 J/cm^2) were evaluated in the previous chapters of the monograph. However, for convenience of comparison, all data are plotted on one graph and are presented in Fig. 5.35.

An analysis of the graph suggests that, regardless of the modification parameters of PMPJ or IPEB (as independent methods), the increase in microhardness is 97% (0.71 GPa). Complex processing, with an electron beam energy of 25 J/cm^2 (modes 1 and

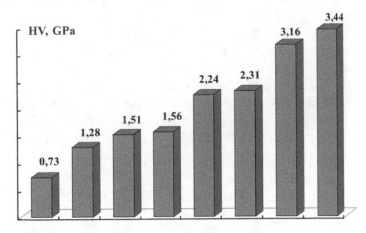

Fig. 5.35. Microhardness of the surface of Al–Si alloy, depending on the modification method. 1 – cast state; 2 – PMPJ (mass of powder sample Y_2O_3 = 58.9 mg, discharge voltage U = 2.8 kV); 3 – PMPJ (Y_2O_3 = 88.3 mg, U = 2.6 kV); 4 – IPEB (25 J/cm^2) 5 – complex processing (mode 1, Table 2.3); 6 – complex processing (mode 2, Table 2.3); 7 – complex processing (mode 3, table 1.1); 8 – complex processing (mode 4. Table 2.3).

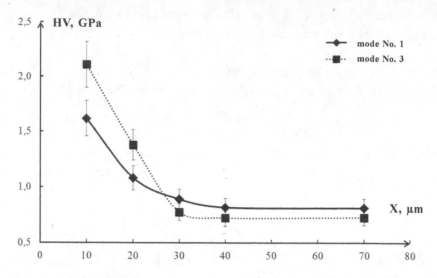

Fig. 5.36. Microhardness values of layers located at different distances (X) from the surface of the complex treated in modes 1 and 3.

2, Table 2.3), leads to an increase in microhardness by ~3.1 times (2.24 GPa and 2.31 GPa, respectively) compared to 0.73 GPa in the cast state. Complex processing, with an electron beam energy of 35 J/cm² (modes 3 and 4, Table 2.3), leads to an increase in microhardness by 4.3 times (3.16 GPa) and 4.7 times (3.44 GPa) depending from the regime, in comparison with 0.73 GPa, it is a molten state, which correlates with the data of tribological tests.

In addition, an assessment was made of the change in microhardness (Fig. 5.36) on the transverse sections, depending on the distance to the processing surface, for the modes 1 and 3.

Analysis of the dependence of the change in microhardness on the distance to the surface of the modification (X) shown in Fig. 5.36, says that regardless of the complex processing mode in a layer located at a depth of 10 µm, the microhardness value is maximum (1.58 ± 0.16 GPa for mode 1 and 2.07 ± 0.21 GPa for mode 3). Further advancement deep into the material leads to a monotonic decrease in microhardness with the achievement of a value characteristic of an alloy not achieved in modification at a distance of 40–50 µm (0.80 ± 0.81 GPa for mode 1 and 0.71 ± 0.70 GPa for mode 3).

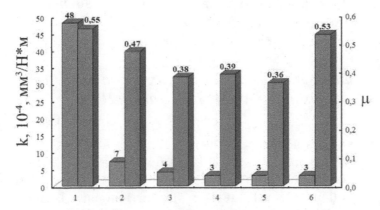

Fig. 5.37. Dependence of the coefficient of friction (μ) and the value of the wear parameter (k) on the state of the samples; 1 – cast state; 2 – state of the material after irradiation (25 J/cm²); 3–6 – state after complex processing, combining exposure to a multiphase plasma jet and irradiation with an intense pulsed electron beam: 3 – mode 1; 4 – mode 2; 5 – mode 3; 6 – mode 4.

5.2.2. Friction tests of Al–Si alloy after electron–ion–plasma treatment

The evaluated frictional properties were the friction coefficient and the wear parameter (value, the reciprocal of wear resistance) and the measurement results are shown in Fig. 5.37.

The tests performed showed that complex surface treatment leads to a multiple increase in the wear resistance of the modified layer and a decrease in the friction coefficient, which is due to the formation of a multiphase submicro-nanocrystalline state.

The wear parameter weakly depends on the modes of combined processing. In relation to the starting material, an increase in wear resistance by (18–20) times was revealed; in the case of the Al–Si alloy irradiated by an intense pulsed electron beam, the increase in wear resistance was (2.6–2.8) times. The friction coefficient changes less unambiguously, namely, it decreases by ≈1.5 times with respect to the initial alloy by 1–3 times and by ≈1.3 times with respect to the alloy irradiated with an intense pulsed electron beam. When combined processing with mode No. 4, the friction coefficient increases, reaching a value close to the value of the friction coefficient of the original Al–Si alloy.

5.3. Physico-mathematical modelling of the formation and evolution of the structure of an Al–Si alloy under electron-ion–plasma exposure

5.3.1. Mathematical model of the formation of surface nanostructures in the Al–Si alloy during IPEB

When studying small-scale regular structures on the surface of absorbing media under the influence of electron beams, it was found that an increase in the electron beam density from $E_s = 10$ J/cm² to $E_s = 30$ J/cm² leads to an increase in cell sizes from (130–670 nm) to (200–800 nm) in rail steel. A similar tendency is observed in Al–Si alloys [7, 8] with E_s varying from 15 J/cm² to $E_s = 35$ J/cm² (Table 5.6).

Table 5.6. Dependence of the size of the structure of cellular crystallization on the energy density of the electron beam

Density of electron beam energy, J/cm²	Cell size in high-speed crystallization, nm
15	200–220
20	400–500
25	215–751
30	350–550

From this table it follows that the cell sizes increase with increasing electron beam energy density, but at $E_s > 20$ J/cm² this increase is rather small.

One of the reasons for the formation of cellular structures can be the appearance and development of various hydrodynamic instabilities [9], such as thermo- and evaporative–capillary instabilities [10, 11]. The mechanism of the formation of cellular structures due to the development of thermocapillary instability was proposed in [12, 13]. Its essence lies in the fact that the presence of a temperature gradient along the depth of the sample leads to the excitation of capillary waves, which increase under the action of the tangent thermocapillary force. This force arises due to the presence of a gradient of surface tension along the longitudinal coordinate. As a result, the amplitude of perturbations of the melt surface increases. This leads to the formation of vortex structures and then cells after crystallization. The growth of cell sizes with increasing E_s, from this point of view, can be explained by the fact that it leads to a decrease in the temperature gradient due to an increase in the penetration depth and, accordingly, an increase in the critical wavelength at which instability begins. A slight change in the cell size at $E_s > 20$ J/cm^2 can be explained by the fact that the evaporation process begins to prevail at these energy densities.

To implement the thermocapillary mechanism of the formation of surface structures with sizes of 200–800 nm, temperature gradients $G_0 \sim 10^9$–10^{10} K/m are required. From the thermal model [13] for Al–Si alloy $G_0 = (T_V - T_L)/h = 10^7$ K/m, where T_V is the evaporation temperature, T_L-, $h \sim 100$ μm is the thickness of the molten layer, this is three orders of magnitude lower than the required values. This discrepancy can be explained by the fact that during the formation of low-energy high-current electron beams, using a plasma diode, a high-voltage discharge is first generated with an action time of ~ 0.1 μs, then the penetration depth is $h \sim 0.1$ μm and $G \sim 10^9$ K/m. In this case, the effect on the surface is a sequence of, two heat fluxes: the first operates one microsecond with a thermal gradient of 10^9 K/m, the second 100 μs and a gradient of 10^7 K/m. Then, in a time of 1 μs, a layer with nanoscale cells is formed near the surface, and conditions for modulating the heat flux along the x coordinate are created. Due to this modulation, elongated cells with a longitudinal size of 80 μm are formed. Another important factor is the evaporation process. Under the conditions of high vacuum, the evaporation temperature of materials according to the Clapeyron–Clausius equation shifts toward low temperatures [14]. The resulting steam, in turn, creates pressure on the surface of the melt, which is called evaporative. The effect of evaporative pressure on the instability of the melt boundary under laser energy influences was studied in [11, 15]. It was shown in [11, 15] that the spatiotemporal modulation of the evaporation pressure, which occurs in the presence of thermal disturbances, makes an additional contribution to their growth and conditions are determined when the influence of the evaporation pressure plays a decisive role. Thus, in order to create mathematical models of the effect of electron beams, it is necessary to take into account the contribution of the evaporative pressure when compiling dynamic conditions at the melt/intrinsic vapour interface.

Thus, the purpose of this section is to study the effect of evaporative pressure on the thermocapillary instability of the molten layer when exposed to low-energy high-current electron beams.

Formulation of the problem

As in [10–13], to solve the problem of the thermocapillary melt flow, we consider a viscous incompressible heat-conducting evaporating liquid. It occupies a layer of thickness h on the free surface $z = \eta(x, t)$, moreover $-\infty < x < +\infty$. Under the influence of an electron beam in the liquid layer, a temperature profile $T_0(z)$ is established, where T_0 is the unperturbed temperature. Due to thermal conductivity, this profile changes over time. However, if the characteristic time of evolution of the perturbations is less than the time of change of $T_0(z)$, then in the analysis of instability the dependence of $T_0(z)$ can be considered unchanged [10, 12]. The melt temperature will consist of the unperturbed temperature $T_0(z)$ and the perturbed $T(x, z, t)$. Let the direction of the wave vector of temperature and velocity perturbations coincide with the direction of the X axis. Then they will depend on the x coordinates and time t according to the law $\exp(\omega t - ikx)$.

To study the instability of a stationary state, we use the linearized Navier–Stokes equations and thermal conductivity. They will look like:

$$\frac{\partial u}{\partial t} = -\frac{1}{\rho}\frac{\partial p}{\partial x} + v\left(\frac{\partial^2 u}{\partial x^2}\right) \quad \frac{\partial w}{\partial t} = -\frac{1}{\rho}\frac{\partial p}{\partial z} + v\left(\frac{\partial^2 w}{\partial x^2} + \frac{\partial^2 w}{\partial x^2}\right),$$

$$\frac{\partial u}{\partial x} + \frac{\partial w}{\partial z} = 0, \quad \frac{\partial T}{\partial t} + wG_0 = \chi\left(\frac{\partial^2 T}{\partial x^2} + \frac{\partial^2 T}{\partial z^2}\right) \tag{5.1}$$

where u, w are the components of the velocity perturbation vector, ρ is the density, p, T are the pressure and temperature perturbations, v, χ, $\sigma(T)$ are the kinematic viscosity, thermal diffusivity, and surface tension, respectively, $G_0 = dT_0/dz$ is the unperturbed gradient temperature along the z axis. We believe that surface tension depends on temperature according to a linear law

$$\sigma = \sigma_0 + \sigma_T (T - T_{00}) \tag{5.2}$$

where σ_T is the temperature coefficient of surface tension ($\sigma_T < 0$), σ_0 is the surface tension at room temperature T_{00}.

The boundary conditions on the surface of the melt $z = 0$ have the form [11, 15]:

$$w = \frac{\partial \eta}{\partial t}, \quad -p + 2\rho v \frac{\partial w}{\partial z} + p_v' T = \sigma_0 \frac{\partial^2 \eta}{\partial x^2},$$

$$\rho v \left(\frac{\partial u}{\partial z} + \frac{\partial w}{dx} \right) = \frac{\partial \sigma}{\partial x}, \quad \frac{\partial T}{\partial x} = 0 \qquad (5.3)$$

where η is the displacement of the melt surface along the z axis, $p_v' = \dfrac{dp_v}{dT}$ is the vapour pressure of the material. At the melt/solid interface $z = -h$, the temperature perturbations and the components of the velocity vector are set equal to zero:

$$u = v = w = 0, \; T = 0. \qquad (5.4)$$

The surface tension gradient along the x axis has the form:

$$\frac{\partial \sigma}{\partial x} = \frac{\partial \sigma}{\partial T} \left(\frac{\partial T}{\partial x} + G_0 \frac{\partial \eta}{\partial x} \right). \qquad (5.5)$$

We seek a solution to system (5.1)–(5.4) in the form

$$u(x,z,t) = U(t)\exp(\omega t - ikx)$$
$$w(x,z,t) = W(z)\exp(\omega t - ikx)$$
$$p(x,z,t) = P(z)\exp(\omega t - ikx)$$
$$T(x,z,t) = T(z)\exp(\omega t - ikx)$$
$$\eta(x,z,t) = \eta_0 \exp(\omega t - ikx). \qquad (5.6)$$

where $U(z)$, $V(z)$, $W(z)$, $P(z)$, $T(z)$, η_0 are the amplitudes of the perturbations of the components of the velocity, pressure, temperature and surface vectors, respectively, k is the wave number, i is the imaginary unit. Substitution of (5.6) into (5.1)–(5.5) leads to the following system:

$$P(z) = \frac{i\rho v}{k} \left(U*(z) - k_1^2(z) \right),$$

$$W''(z) - k_1^2 W(z) = \frac{P'(z)}{\rho v},$$

$$U(z) = -\frac{i}{k} W'(z), \qquad (5.7)$$

$$T''(z) - k_2^2 T - \frac{G_0}{\chi} W = 0,$$

where $k_1^2 = k^2 + \dfrac{\omega}{v}$, $k_2^2 = k^2 + \dfrac{\omega}{\chi}$, . A prime means differentiation with respect to z. After the conversion, the first and second equations of system (5.7) take the form:

$$W^{IV}(z) - \left(k^2 + k_1^2\right)W''(z) + k^2 k_1^2 W(z) = 0,$$

$$T''(z) - k_2^2 T - \frac{G_0}{\chi}W = 0. \tag{5.8}$$

The boundary conditions (5.3), (5.4), taking into account (5.6) and (5.7), will have the following form:

$$W'''(0) - (k_1^2 + 2k^2)W'(0) - k^3\left(\frac{W(0)\omega_c^2}{\omega\omega_v} + \frac{T(0)\omega_p}{G_0}\right) = 0$$

$$W''(0) + k^2 W(0) + \frac{\omega_T k^2 \Sigma_1(0)}{\omega} = 0 \tag{5.9}$$

$$T'(0) = 0, \ W(-h) = W'(-h) = T(-h) = 0,$$

where $\omega_c^2 = \dfrac{\sigma_0 k^2}{\rho}$, $\omega_T = \dfrac{\sigma_2 G_0}{\rho v}$, $\omega_v = vk^2$, $\Sigma_1(0) = \dfrac{\omega}{G_0}T(0)$, $\omega_p = \dfrac{p_v' G_0}{\rho vk}$.

We assume that $h \to -\infty$. The solution (5.8), which satisfies the boundary conditions on $z \to -\infty$, has the form:

$$W(z) = A_1 \exp(kz) + A_2 \exp(k_1 z)$$

$$T(z) = C\exp(k_2 z) - \frac{G_0}{\omega}\left(A_1 \exp(kz) - A_2 \delta \exp(k_1 z)\right); \delta = \frac{\varepsilon}{1-\varepsilon}, \varepsilon = \frac{v}{\chi} \tag{5.10}$$

where A_1, A_2, C are the integration constants. Define Σ_1 through the coefficients A_1 and A_2

$$\Sigma_1 = \frac{\omega}{G_0}T(0) = \frac{1}{k_2}\left(A_1(k - k_2) + \delta A_2(k_2 - k_1)\right) = \left(\frac{k}{k_2} - 1\right)A_1 + \delta\left(1 - \frac{k_1}{k_2}\right)A_2$$

$$\tag{5.11}$$

Substitution of (5.10) and (5.11) into the first and second equations (5.9) allows the system of equations for A_1 and A_2:

$$a_{11}A_1 + a_{12}A_2 = 0$$
$$a_{21}A_1 + a_{22}A_2 = 0 \tag{5.12}$$

where

$$a_{11} = -\frac{k^3(k-k_2)\omega_p}{\omega k_2} - k(k^2 + k_1^2) - \frac{k^3\omega_c^2}{\omega\omega_v}, \quad a_{12} = \frac{k^3\delta(k_1-k_2)\omega_p}{k_2\omega} - 2k^2 k_1 - \frac{k^3\omega_c^2}{\omega\omega_v},$$

$$a_{21} = 2 - \frac{(k-k_2)\omega_T}{\omega k_2}, \quad a_{22} = 1 + \frac{k_1^2}{k^2} + \frac{\delta(k_1-k_2)\omega_T}{\omega k_2}.$$

The determinant of this system will be the dispersion equation. We represent this equation in the form

$$R_\sigma - R_T - R_V = 0, \tag{5.13}$$

where

$$R_\sigma = \omega^2\left((\omega + 2\omega_V)^2 + \omega_c^2 - 4z_1\omega_v^2\right),$$

$$R_T = \omega_v\omega_T\left(\begin{array}{c}(1-z_1/z_2)\delta(\omega^2 + 2\omega\omega_v + \omega_c^2) \\ +(1-1/z_2)(2\omega\omega_v z_1 + \omega_c^2) + \omega(\omega + 2\omega_v(1-z_1))\end{array} + \right)$$

$$R_V = \left(2\left(\delta\left(1-\frac{z_1}{z_2}\right) + \left(1-\frac{1}{z_2}\right)\right)\omega\omega_v^2 + \left(1-\frac{1}{z_2}\right)\omega^2\omega_v\right)\omega_p. \tag{5.14}$$

We will consider the approximation, moreover, for liquid metals. Therefore, we can write approximate relations, neglecting:

$$\frac{1}{z_2} \approx 1 - \frac{\varepsilon\omega}{2\omega_v} \Rightarrow 1 - 1/z_2 \approx \varepsilon\omega/2\omega_v$$

$$\text{and } (1-z_1/z_2)\delta \approx (1-z+z\varepsilon\omega/2\omega_v)\varepsilon \approx (1-z_1)\varepsilon. \tag{5.15}$$

Using (5.15) and substitution, equation (5.13) takes the form:

$$(z_1+1)^2\left((z_1^2+1)^2 + C^2 - 4z_1\right) - C_1(2(z_1^2-1)C^2) =$$

$$-C_2(z_1^2 + 2z_1 + 3)(z_1^2 - 1) = 0, \tag{5.16}$$

where $C^2 = \frac{\omega_c^2}{\omega_v^r}, \quad C_1 = \frac{\omega_T\varepsilon}{2\omega_v}, \quad C_2 = \frac{\omega_p\varepsilon}{2\omega_v}.$

Results and discussion
We find the initial wavelength of perturbations of the melt boundary

at which instability occurs. To do this, we will seek a solution to (5.16) in the form $z^2 = z_0^2 + \tilde{\omega}$, $z_0^2 = -1 \pm iC$, $|\tilde{\omega}| \ll |z_0^2|$. As a result, we obtain corrections to the frequency of capillary waves due to the presence of a temperature gradient and evaporative pressure:

$$\tilde{\omega} = \frac{C_1}{2} + \frac{C_2}{2}.$$ (5.17)

Given that we get

$$\omega = -2\omega_v = \frac{\omega_T \varepsilon}{4} + \frac{\omega_T \varepsilon}{4} + i\omega_c.$$ (5.18)

Capillary waves will be unstable (Re $\omega > 0$, where Re is the real part of the complex number $\omega = \alpha + i\Omega$, where α is the growth rate of perturbations of the melt surface, Ω is their cyclic frequency) if, $\omega_T < 0$. $\omega_p > 0$ and $\left(|\omega_T| + \omega_p\right) > \frac{8}{\varepsilon}\omega_v$. The contribution of the evaporative pressure becomes decisive for the instability of capillary waves under the condition $k_p \le -\frac{p_v'}{\sigma_T}$ [104]. Since, then this instability occurs when, that is, with a decrease in temperature deep into the molten layer. Substitution of the numerical data for the Al–Si alloy

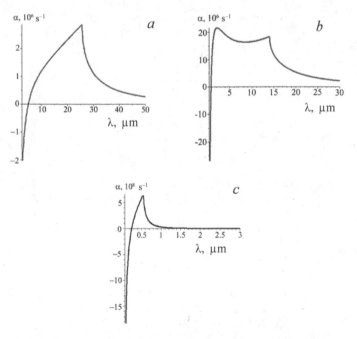

Fig. 5.38. Dependence of the growth rate on wavelength. *a)* $p_v'(T_V-T_L) = 2 \cdot 10^5$ Pa and $h \sim 10^{-6}$ m; *b)* $p_v'(T_V-T_L) = 2 \cdot 10^4$ Pa and $h \sim 10^{-8}$ m and *c)* $p_v'(T_V-T_L) = 2 \cdot 10^{11}$ Pa and $G_0 \sim 10^7$ K/m.

The size distribution of cells is single-mode. Its maximum occurs at 0.415 μm. Perturbations with a wavelength of this range according to (5.18) are possible at a temperature gradient of $\sim 10^{11}-10^{12}$ K/m

Table 5.7. Thermophysical characteristics of an Al–Si alloy (12% Si, Al – the rest)

Notation/dimension	Al–Si alloy	Characteristic
T_L, K	850	Melting point
T_V, K	1270	Melting point
ρ_c, kg/m³	2656	Solid phase density
ρ_L, kg/m³	2398	Liquid phase density
v, 10^{-7}, m²/s	3.5	Viscosity
χ, 10^{-7}, m²/s	3.3	Heat conductivity
σ, N/m	0.87	Surface melting
σ_T, 10^{-7}, N/(m N)	−0.35	Temperature coefficient of surface melting

(Table 5.7.) Shows that perturbations with a wavelength of $\lambda > 5.21$ μm are unstable for the thickness of the molten layer $h \sim 10^{-6}$ m and $p_v'(T_V - T_L) = 2 \cdot 10^5$ Pa. This is confirmed by a numerical solution of equation (5.16). From Fig. 5.38 a it is seen that the value of the growth rate of perturbations of the melt surface is $\alpha \approx 0$ at $\lambda = 5.2$ μm.

At $\lambda > 5.2$ μm, perturbations increase due to thermocapillarity and evaporation, and the contribution of evaporative pressure is decisive, since $\lambda_p = 2\pi/k_p \approx 4.62$ μm. The maximum growth rate of disturbances is achieved at $\lambda = 25.7$ μm. At $h \sim 10^{-8}$ m and $p_v'(T_V - T_L) = 2 \cdot 10^4$ Pa, the critical wavelength is $\lambda_c = 0.75$ μm. The maximum growth rate (Fig. 1b) occurs at wavelengths of 2.25 μm and 14.09 μm. The first maximum is due to thermocapillary phenomena, and the second is due to evaporative pressure. Note that in [13, 15] the two-mode dependence of the growth rate on the wavelength was obtained from the solution of equation (5.14) at $R_V = 0$. Indeed, at a maximum growth rate occurs at a wavelength of 2.25 μm, and a second maximum does not occur. Thus, the evaporative pressure significantly changes the dependence of the growth rate of disturbances on the wavelength.

The experimental data on the sizes of crystallization cells (Fig. 5.39) [7, 8] show that their sizes are in the range from 0.215 μm to 0.751 μm.

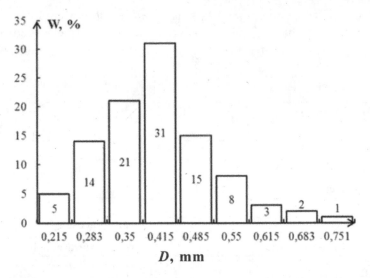

Fig. 5.39. Size distribution of cells of high-speed crystallization of the surface layer by size for material irradiated by an intense pulsed electron beam.

and $p_v'(T_V-T_L)=2\cdot10^4$ Pa or at $G_0 \sim 10^7$ K/m and $p_v'(T_V-T_L)=2\cdot10^{11}$ Pa (Fig. 5.38 c). This is confirmed by a numerical solution of equation (5.16). The above facts allow us to conclude that the applied low-frequency approximation provides an adequate explanation for the single-mode distribution of high-speed crystallization cells during IPEB, but does not explain the appearance of crystallization cells larger than 0.1 µm at $G_0 <10^{11}$ K/m. Therefore, we use the complete dispersion equation (5.14). Using the substitution $\omega = \omega_v(z_1^2-1)$ and subsequent transformations, this equation is reduced to an algebraic equation of the 16th degree, which, due to its cumbersome nature, will not be written out. In this case, solutions that satisfy the conditions Re $(\omega) > 0$, Re $(z_1) > 0$, Re $(z_2) > 0$ will be unstable. The results of the numerical solution for the temperature gradient $G_0 = 8.4 \cdot 10^9$ K/m and Pa are presented in Fig. 5.40. As without taking into account evaporation effects [16], equation (5.14) admits two unstable solutions, which indicates the existence of two dependences of the growth rate on the wavelength. The first dependence (Fig. 5.40 a) has a maximum at $\lambda = 0.5$ µm, with $\alpha_m \approx 10^9$ s^{-1}, which practically coincides with the experimental data [17]. The second dependence also has one maximum at $\lambda = 5.1$ µm, with $\alpha_m \approx 5 \cdot 10^6$ s^{-1} (Fig. 5.40 b). Apparently, it gives an explanation for the formation of surface periodic structures observed in experiments [105, 106].

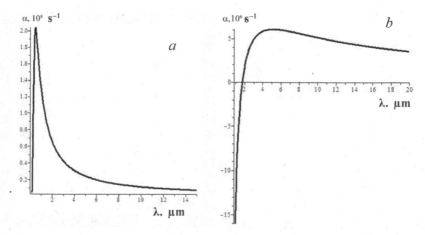

Fig. 5.40. Dependences of the growth rate on the wavelength at $p_v'(T_V-T_L) = 2\cdot10^4$ Pa and $G_0 = 8.4 \cdot 10^9$ K/m, obtained by solving equation (5.14). *a* is the first root of equation (5.14), *b* is the second root of equation (5.14).

A theoretical study of the formation of a cellular structure of high-speed crystallization under the influence of low-energy high-current electron beams on Al–Si alloys is carried out. A mathematical model of the formation of these structures is proposed, based on the concept of the occurrence of thermocapillary instability of the "liquid phase / intrinsic vapour" interface and taking into account the vaporization pressure. It is shown that taking vapour pressure into account in the low-frequency approximation leads to a two-mode dependence of the growth rate on the wavelength at Pa and a temperature gradient of $4.2 \cdot 10^{11}$ K/m. The solution of the complete equation (5.14) showed that, regardless of whether the evaporative pressure is taken into account or not, there are two dependences of the velocity on the wavelength. The first dependence has one maximum at a wavelength of 500 nm. This allows us to conclude that it is responsible for the formation of a cellular structure on the surface. The second dependence has one maximum in the microdimensional range. This dependence, apparently, is responsible for the formation of ordered surface relief structures.

5.3.2. Physico-mathematical model of the evolution of the structure of an Al–Si alloy modified with yttrium when exposed to an electron beam and fuel combustion products

The disadvantage of Al–Si alloys is the low mechanical properties that arise due to the formation of coarse dispersed inclusions of silicon. These inclusions are concentrators of mechanical stresses, the high level of which leads to the formation of cracks. Therefore, it is necessary to develop methods for reducing the size of silicon inclusions. Currently, various types of heat treatment are used to solve this problem, including the use of concentrated energy flows (laser processing, electron-beam processing, etc.). During the heat treatment of multiphase alloys, particles of the second phase undergo two types of transformations: 1) coalescence, which consists in the enlargement of these particles; 2) crushing of particles of the second phase with subsequent spheroidization. Let us dwell in more detail on the second type of transformation. According to [18, 19], this process occurs according to the diffusion mechanism due to the concentration gradient at the 'second phase/matrix' interface. With increasing temperature, this process accelerates. The precipitates of the second phase can acquire equiaxial shape by dividing the plates or needles into several particles. An important role in fission belongs to defects in the crystal lattice of the matrix and the second phase. Works [20, 21] show that the formation of equiaxed particles of the second phase is due to the influence of surface tension and a simple dynamic model of this process is proposed. In the works of M.A. Grinfeld [22, 23], by analyzing the second variation of the free energy functional of the 'melt/crystal' system, it was found that the non-hydrostatic components of the stress field inside the elastic crystal lead to instability of their interface, while for the manifestation of this instability, the liquid could dissolve the solid phase or contribute to particle transport along the crystal boundary. Interfacial surface tension cannot suppress this instability in the long-wavelength region of the spectrum, although it has a stabilizing effect. This instability is also manifested if both phases are solids. The main condition for its occurrence in this case is that the shear stresses exceed a certain critical value, which in turn depends on the ratio of the longitudinal and transverse sound velocities.

In [24], the behaviour of a pearlitic structure under pulsed loading was analyzed. It is shown that at the initial stage of high-speed tension (in the zone of interference of unloading waves), the fragmentation of pearlite components (cementite and ferrite) to ultrafine sizes occurs. Cementite, as an unstable phase begins to decompose with the formation of carbon, which interacts with ferrite, and at the ferrite–cementite interface, new cementite globules nucleate in some places. The second stage of the mechanochemical process of pearlite spheroidization is due to the additional supply of carbon atoms from the matrix to the zone of spall damage. Ultrafine particle size, the dissolution of pearlite components and their enrichment with carbon in a repeated chemical reaction can lead to an increase in the amount of cementite due to the chemical reaction of the interaction of additional carbon with iron. A similar mechanism was found in differentially hardened rail steel during prolonged operation [25, 26]. The dissolution of carbon particles in titanium under the action of electron beams was studied in [27, 28], where it was shown that the main mechanism of dissolution is diffusion and the dependence of the inclusion thickness on time is obtained. It has been established that nanosized particles dissolve faster than microdimensional ones. For an Al–Si alloy, the diffusion coefficient of silicon into aluminium is about 10–16 cm^2/s; accordingly, the dissolution time of inclusions reaches ~10–100 s, which allows us to conclude that in aluminium-silicon alloys, diffusion does not diffuse is the main mechanism of crushing and spheroidization. In [29], a mechanism was proposed, a mechanism was proposed according to which spheroidization occurs due to the destruction of silicon wafers due to the difference in the coefficients of linear thermal expansion of the matrix and inclusion. Since, compared with the aluminium matrix, the volume fraction of silicon wafers is small, the aluminium matrix makes the main contribution to thermal expansion. The linear expansion coefficient of aluminium is 4 times that of silicon. In this regard, the thermal expansion (compression) of the two phases is in most cases incompatible. This leads to the inevitable occurrence of mechanical stresses between the phases. Silicon inclusions are able to accept only ¼ of the thermal expansion (compression) transmitted by the aluminium matrix, through its own thermal expansion (compression). The rest goes to the deformation of the matrix and the destruction of the silicon wafers (due to their brittleness). The occurrence of cracks is due to inhomogeneities of the inclusion surface. The resulting cracks will be capillaries for aluminium atoms. The mechanical stresses generated by the cracks will be analogues of capillary forces that move the atoms of the matrix into the formed gaps between the inclusions. In the opposite direction, there are flows of vacancies and silicon atoms. A similar mechanism for crushing particles of the second phase was established upon exposure to high-density pulsed electric current [30] and selective laser melting [31]. In [32, 33], the instability of the interface between materials under the action of contact load

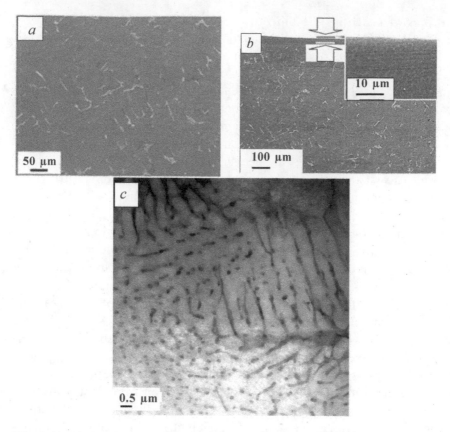

Fig. 5.41. Al–Si alloy structure: *a*) initial state; *b*) and *c*) after irradiation with a pulsed electron beam with an energy density of 35 J/cm² [34, 35].

was studied. Linear analysis showed that there are two classes of new instabilities that are qualitatively different from waveguide instabilities: 1) dynamic instability due to modes propagating with the speed of the dilatation wave in the opposite direction to the slip motion with a small wave number; 2) dynamic instability, which arises due to modes propagating with shear wave velocity in the sliding direction.

Thus, despite the difference in the mechanisms of crushing, it is common that they lead to instability of the surface of the large inclusion of the second phase.

Figure 5.41 *a* shows an electron microscopic image of the structure of the investigated material in the initial state.

From this figure it follows that the material is a multiphase aggregate, the structure of which is represented by grains of an aluminium-based solid solution, Al–Si eutectic grains, inclusions of primary silicon and intermetallic compounds, the sizes and shape of which vary over a very wide range. Electron beam irradiation leads to the formation of a multilayer gradient structure. According to the morphology of the defective substructure, three layers can be conditionally distinguished, which in [35] of this work are called the surface, transition, and heat-affected layers. The surface layer has a columnar crystallization structure formed during high-speed cooling of the material from the molten state (Fig. 5.41 *b*). The thickness of this layer according to scanning electron microscopy is from 70 to 100 μm. A detailed analysis of the transition layer by transmission electron microscopy [34] showed that it is represented by primary inclusions of the second phase (Fig. 5.41 *c*), which are the centers of crystallization of aluminium. The size distribution of these inclusions is bimodal. The average particle size is 138.9 ± 45.3 nm [34].

We assume that with IPEB the following mechanism of crushing of silicon wafers takes place in the heat-affected zone. Consider a silicon wafer enclosed in an aluminium matrix. As already mentioned in the Introduction, due to the mismatch between the elastic moduli and the linear expansion coefficients of aluminium and silicon, mechanical stresses arise on the boundary of the inclusion and matrix, which lead to its instability and destruction. This mechanism can also occur during heating and subsequent cooling of the product by the products of fuel combustion. Fig. 5.42 shows that in the zone of thermal influence of the electron beam, temperatures reach values from 700 to 800 K, the same temperatures are observed in products from Al–Si alloys when exposed to combustion products.

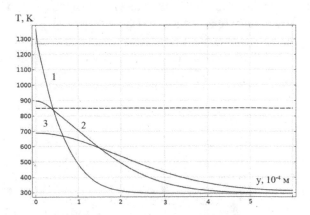

Fig. 5.42. The dependence of the temperature of the yttrium-modified Al–Si alloy on the coordinate at different time points during IPEB: 1 – 150 μs; 2 – 550 μs; 3 – 1000 μs.

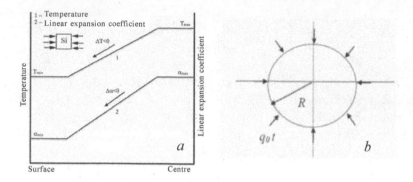

Fig. 5.43. Scheme of interaction of a silicon wafer with an aluminum matrix. *a*) a change in temperature (curve 1) and linear expansion coefficient (curve 2) at the cooling stage; *b*) loading diagram of a silicon wafer.

In Fig. 5.43 is a diagram of the interaction of a silicon wafer and an aluminium matrix in the cooling stage. At this stage, the silicon wafer according to [123] will be loaded with compressive stresses. We approximate the inclusion form by a circle and apply the representations of the theory of stability of elastic systems [36 -39]. Based on these ideas, we consider two cases: 1) the inclusion of silicon is a plate articulated on all sides; 2) the plate is pinched on all sides. In all cases, the force P is applied to the plate along the radius. To determine this, it is necessary to solve the problem of the stress–strain state near the inclusion. To do this, we write the system of differential equations for radial displacement ($u(r)$) at constant values of the elastic modulus (E), Poisson's ratio (v) and linear expansion coefficient (α):

$$\frac{d}{dr}\left(\frac{1}{r}\frac{dru_n}{dr}\right) = (1+v_n)\alpha_n\frac{dT_n}{dr}. \tag{5.19}$$

The first layer is in the range $0 < r < a$ and is characterized by the parameters of the material E_1, v_1, α_1, the second is $a < r < \infty$ and E_2, v_2, α_2, where E_n, v_n, α_n are the elastic modulus, Poisson's ratio, and linear expansion coefficient of the nth layer, respectively. External boundary conditions for $r = 0$ and $r \to \infty$:

$$u_1 = 0, \; u_2(\infty) = 0. \tag{5.20}$$

A solution satisfying external boundary conditions will take the form:

$$u_1(r) = (1+v_1)\frac{\alpha_1}{r}\int_0^r T_1(\xi)\xi d\xi + C_1 r,$$

$$u_2(r) = (1+v_2)\frac{\alpha_2}{r}\int_0^r T_2(\xi)\xi d\xi + \frac{C_2}{2}. \tag{5.21}$$

Stress components

$$\sigma_{r1}(r) = \frac{\alpha_1 E_1}{r^2}\int_0^r T_1(\xi)\xi d\xi + C_1\frac{E_1}{1-v_1},$$

$$\sigma_{r2}(r) = \frac{\alpha_2 E_2}{r^2}\int_0^r T_2(\xi)\xi d\xi - C_2\frac{E_2}{r^2(1-v_2)}, \tag{5.22}$$

We write the conjugation conditions between the layers in the form of the equality of radial stresses and displacements at the contact point of the layers $r = a$:

$$-E_1\Phi + C_1\frac{E_1}{1-v_1} = -C_2\frac{E_2}{a^2(1+v_2)}, \quad (1+v_1)a\Phi + C_1 a = \frac{C_2}{a};$$

$$\Phi = \frac{\alpha_1}{a^2}\int_0^{\alpha_1} T_1(\xi)\xi d\xi. \tag{5.23}$$

The system (5.23) solution

$$C_1 = \left(\frac{1-K}{1+K} - v_1(1+K)\right)\Phi, \quad C_2 = \frac{2a^2\Phi}{1+K}; \quad K = \frac{E_2(1-v_1)}{E_1(1+v_2)}. \tag{5.24}$$

To calculate stresses, it is necessary to determine the temperature distribution. Consider the case when the temperature is constant, then

$$\Phi = \frac{\alpha_1}{2}T_0. \tag{5.25}$$

Taking into account (5.24) and (5.25), the stress at the inclusion boundary will take the form:

$$\sigma_1 = \frac{B_1 B_2(1+v_1)T_0\left((\alpha_1 = \alpha_2)v_2 - (\alpha_1 + \alpha_2)\right)}{B_1 v_1 - B_2 v_2 + B_1 + B_2}, \tag{5.26}$$

where $B_n = \dfrac{E_n}{1-v_n}$, $n = 1...2..$

After searching for the distribution of stresses along the inclusion radius, we turn to the consideration of its stability. The basic equation of motion in terms of the theory of plates and shells [38, 39] will have the form:

$$D\Delta\Delta w + P\Delta w + \rho h \frac{\partial^2 w}{\partial t^2} = 0, \qquad (5.27)$$

where w is the transverse displacement, P is the load, ρ is the density of the plate material, h is its thickness, $D = \dfrac{Eh^3}{12\left(1-v^2\right)}$ is the bending stiffness, Δ is the Laplace operator in polar coordinates. We will seek a solution to (5.27) in the form

$$w(r, \varphi, t) = Z(t)w_{mn}(r) \cos n\varphi \qquad (5.28)$$

where $Z(t)$ and $w_{mn}(r) \cos (n\varphi)$ are the temporal and coordinate part of the transverse displacement. The amplitude of this displacement $w_{mn}(r)$ depends on the form of the boundary conditions. In case of articulation, the boundary conditions for equation (5.27) have the form:

$$\frac{d^2 w_{mn}}{dr^2} + \frac{v}{r}\frac{dw_{mn}}{dr} = 0, \text{ at } r = R, \qquad (5.29)$$

and when pinched:

$$w_{mn} = \frac{dw_{mn}}{dr} \text{ at } r = R. \qquad (5.30)$$

The initial conditions will be:

$$Z(0) = Z, \dot{Z}(0) = 0. \qquad (5.31)$$

We will use the instability criterion proposed in [38]:
$$Z(t_{cr}) = 0, \qquad (5.32)$$
where t_{cr} is the critical time (the moment of instability onset).

Results and discussion

To solve problem (5.27)–(5.32) we use the approximate Bubnov-Galerkin method. We set the external load in the form, where q_0 is the loading speed. In the case of a pivotally supported plate, the coordinate part of its transverse displacement has the form:

$$w_{mn}(r) = J_n\left(\beta_{n,m}R\right)\left(\frac{r}{R} - \left(\frac{r}{R}\right)^n\right),$$ (5.33)

where $\beta_{n,m}$ is the root of the equation:

$$\left(\beta_{n,m}R\right)J_n\left(\beta_{n,m}R\right) - (1-v)J_{n+1}\left(\beta_{n,m}R\right) = 0$$ (5.34)

where $J_n(\beta_{n,m}, R)$ is the nth order Bessel function. The critical load in this case has the form:

$$q_0 t_{cr} = \beta_{n,m}^2 \frac{D}{R^2} + \frac{b_1\rho h R^2}{\beta_{m,n}^2 t_{cr}^2} \frac{\left(\beta_{m,n}^2 - 2(1+v)\right)}{\beta_{m,n}^2 - 1 - v^2}.$$ (5.35)

For a jammed plate

$$w_{mn}(r) = J_n\left(\alpha_{n+1,m}R\right)\left(\frac{r}{R} - \left(\frac{r}{R}\right)^2\right),$$ (5.36)

where $\alpha_{n+1,m}$ is the root of the equation:
$$J_n(\alpha_{n+1,m}R) = 0$$ (5.37)
The value of the critical load $P_{cr} = q_0 t_{cr}$ in the case of a pinched plate has the form:

$$q_0 t_{cr} = \alpha_{m,n}^2 \frac{D}{R^2 h} + \frac{b_1\rho R^2}{\alpha_{m,n}^2 t_{cr}^2}.$$ (5.38)

In the case of static loading for a pivotally supported plate, the critical load value will take the form:

$$P_{cr} = \beta_{m,n}^2 \frac{D}{R^2 h},$$ (5.39)

and for a pinched plate:

Table 5.8. Characteristics of matrix material and inclusion

Characteristic	Material	
	Al	Si
Elasticity modulus, GPa	70	110
Linear expansion coefficient, K^{-1}	$28.1 \cdot 10^{-6}$	$3.68 \cdot 10^{-6}$
Poisson coefficient	0.3	0.3

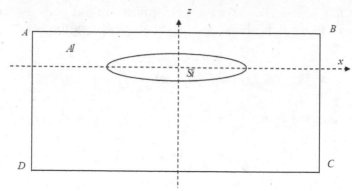

Fig. 5.44. Statement of the problem of calculating stresses near an elliptical inclusion

$$p_{cr} = \alpha_{m,n}^2 \frac{D}{R^2 h}.$$

(5.40)

For large roots $\alpha_{m,n} = \frac{\pi}{4}(2m + 4n + 1)$, $\beta_{m,n} = \pi\left(2n + m + \frac{3}{4}\right)$. Table 5.8 presents the initial data for the calculations.

The critical loads according to (5.39) and (5.40) for switching on with a radius of $R \sim 10$ μm and a thickness of $h \sim 1$ μm are: $p_{cr} \approx 4.22 \cdot 10^8$ Pa and $p_{cr} \approx 1.48 \cdot 10^8$ Pa. According to (8), the stress at the inclusion interface at a temperature of 577 K is 1.19 GPa. For inclusion with a thickness of $h \sim 100$ nm, the critical loads are: for the first case, $p_{cr} \approx 4.25 \cdot 10^6$ Pa, for the second case, $p_{cr} \approx 1.49 \cdot 10^8$ Pa.

To find thermoelastic stresses near the inclusions of the second phase of an elliptical shape, we apply the finite element method. The problem statement for this element is shown in Fig. 5.44. The process of heating and cooling the material will be simulated both with IPEB and under the conditions of exposure to combustion products during operation.

The equations of the theory of thermoelasticity will have the form:

$$\rho \frac{\partial^2 \mathbf{u}_s}{\partial t^2} = \nabla \sigma + \mathbf{F}$$

$$\sigma = C : \varepsilon$$

$$\varepsilon = \frac{1}{2}\left[\nabla \mathbf{u} + (\nabla \mathbf{u}_s)^T\right] + \varepsilon_T$$

(5.41)

$$\varepsilon_T = \alpha(T - T_0),$$

Table 5.9. Characteristics of matrix material and inclusion

	T, q	$\sigma, \mathbf{u}$
AB	$T=T_{in}(t)$	
AD, BC	$T_{AD} = T_{BC}$	$\mathbf{u}_{AD} = \mathbf{u}_{BC}$
DC	$-\mathbf{n}\, q = 0$	$\mathbf{u} = 0$

where $\mathbf{u}_x = (u_s,\, v_s,\, w_s)$ is the displacement vector, respectively, in the $(r,\, \varphi,\, z)$ coordinates, σ is the stress tensor, ε is the deformation tensor, $\mathbf{C} = C(E,\, \mu)$ is the stiffness tensor, depending on the Young's modulus (E) and Poisson's ratio (μ), α is the coefficient of thermal expansion. The temperature field is calculated using the Fourier heat equation:

$$\rho C_p \frac{\partial T}{\partial t} = \nabla(k\nabla T), \qquad (5.42)$$

where T is the temperature, C_p is the specific heat, k is the coefficient of thermal conductivity. The boundary conditions are given in Table 5.9.

The transition to the plastic region was characterized by the Mises criterion:

$$\sqrt{\left(\sigma_{xx} - \sigma_{yy}\right)^2 + 4\tau_{xy}} \le \sigma_y, \qquad (5.43)$$

where σ_{xx} and σ_{yy} are normal components of the stress tensor, τ_{xy} is the tangent component, and σ_y is the yield strength. Figure 5.45 shows the temperature distribution over the depth of the product at different points in time. It can be seen that at $t < 550$ μs the temperature of switching on and the matrix increases (Fig. 5.45 a, b) to 780 K, and then they cool down (Fig. 5.45 c) to a temperature of 560 K. The temperature front moves from the surface to the depth of the product, heating the inclusions of the second phases in its path. The time dependence of the temperature in the inclusion center is shown in Fig. 5.46. It follows from this figure that the temperature maximum occurs at a time instant of 550 μs and amounts to 780 K. As mentioned above, the linear expansion coefficient of aluminium is 4 times greater than that of silicon, therefore, compressive

224

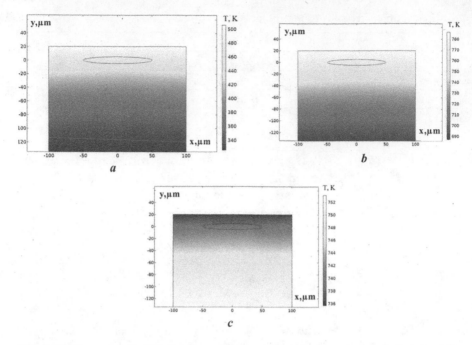

a

b

c

Fig. 5.45. Temperature distribution over the product depth at various points in time (*a*) 150 μs; *b*) 550 μs; *c*) 1000 μs).

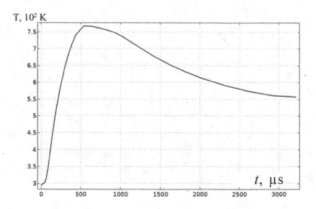

Fig. 5.46. Dependence of temperature on time in the centre of a silicon grain.

mechanical stresses will arise at the interface between silicon and aluminium will lead to its destruction. Based on this, one should expect a maximum of these stresses at a time instant of 550 μs. As the results of calculating the distribution of the Mises stress intensity along the coordinate (Fig. 5.47) show, that at the stage of increasing

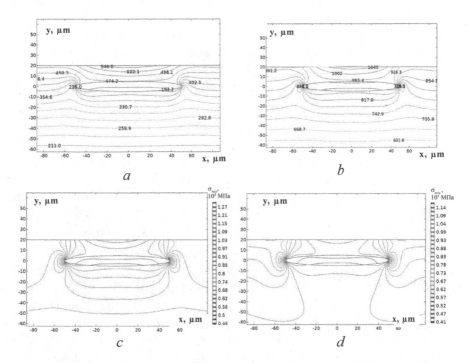

Fig. 5.47. Distribution of stress intensity (MPa) along the coordinate *a*) 150 µs; *b*) 300 µs; *c*) 550 µs; *d*) 1000 µs).

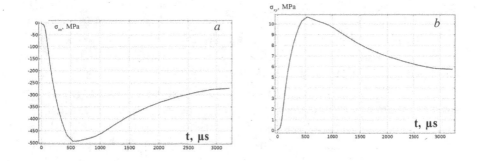

Fig. 5.48. Dependence of the normal components of the stress tensor on time at the inclusion boundary a) the dependence of σ_{xx} on time at the point (0.5L, 0); b) the dependence of σ_y on time at the point (0, 0.5H).

temperature, the stress intensity in the matrix and on (Fig. 5.47a, b) increases. At *t* = 550 µs, the stress intensity reaches its maximum, and in the aluminium matrix, their value is 1.27 GPa, and in the inclusion it is 560 MPa (Fig. 5.47 *c*). At the cooling stage the stress intensity decreases (Fig. 5.47 *d*). The normal stress components at the inclusion boundary behave in the same way (Figs. 5.48 *a* and *b*). The maximum value of σ_{xx} reaches -560 MPa at the point $(0.5L, 0)$, and $\sigma_{yy} = +15$ MPa at the point $(0, 0.5H)$, where L is the longitudinal dimension of the inclusion, H is the transverse dimension of the inclusion. This behaviour of the mechanical stresses allows us to conclude that, for an elliptical inclusion, elastic instability also takes place, which leads to its destruction.

Thus, a mechanism is proposed for the decay of particles of the second phase in an Al–Si alloy in the heat-affected zone of a low-energy high-current electron beam and when exposed to combustion products during operation. This inclusion was modelled by a round plate of radius R and thickness h, as well as an elliptical plate of longitudinal size L and transverse size H. It was assumed that the silicon inclusion is decomposed due to the mismatch between the elastic modulus and the thermal coefficient of linear expansion due to the development of dynamic instability. Under conditions of high-speed cooling, the silicon wafer is loaded with compressive stresses, since the linear expansion coefficient of aluminium is greater than silicon. Under the influence of these forces, instability of the plate occurs and is followed by plate destruction. Estimation of the magnitude of these stresses by the methods of elasticity theory for circular inclusions showed that at an eutectic temperature it can reach about 1 GPa, and for an elliptical inclusioon from 560 MPa to 1.27 GPa. The initial stage of this instability was studied using methods of the theory of plates and shells. In the framework of this theory, the value of the critical stress in the approximation of the clamped ends of the plate and articulated supports is estimated to be $\sim 10^7 - 10^9$ Pa, which coincides with calculations according to the theory of elasticity. This allows us to conclude that the established mechanism of crushing of silicon particles in the zone of thermal influence of the electron beam is the most probable [40].

5.4. Conclusions for Chapter 5

1. A complex surface treatment of the Al–Si alloy was carried out, combining PMPJ with titanium and yttrium oxide and subsequent irradiation with an intense pulsed electron beam;

2. Studies performed by methods of modern physical materials science have revealed the formation of an extended surface layer, the concentration of titanium and yttrium in which depends on the PMPJ regime, and on the distance to the surface of the modification;

3. The mapping method revealed an inhomogeneous distribution of atoms of alloying elements in the modified layer;

4. It was found that PMPJ is accompanied by both doping of the surface layer with plasma elements and the introduction of yttrium oxide powder into the surface layer of particles;

5. Complex processing leads to the formation of a multiphase submicron-nanoscale state in the surface layer of the material, the crystallite sizes of which vary from a few to hundreds of nanometers;

6. It is shown that the wear resistance of the material weakly depends on the option of combined processing. In relation to the initial state of the material, an increase in wear resistance by (18-20) times was revealed; in relation to the Al–Si alloy irradiated by an intense pulsed electron beam, the increase in wear resistance was (2.6–2.8) times;

7. It was revealed that the friction coefficient decreases under the combined treatment modes No. 1–3 in relation to the initial state of the material by ≈1.5 times and in relation to the Al–Si alloy irradiated by an intense pulsed electron beam, by ≈1.3 times. When combined processing according to mode No. 4, the coefficient of friction of the modified layer is close to the value of the coefficient of friction of the initial Al–Si alloy;

8. It was found that complex processing, with an electron beam energy of 25 J/cm², leads to an increase in microhardness by ~3.1 times. Complex processing, in an electron beam energy of 35 J/cm², leads to an increase in microhardness by 4.3 times and 4.7 times, depending on the mode of PMPJ.

9. A mathematical model has been developed for the formation of a cellular structure of high-speed crystallization, based on the concept of the occurrence of thermocapillary instability of the "liquid phase /intrinsic vapour" interface and taking into account the vaporization pressure.

10. A mechanism is proposed for the decay of second-phase particles in an Al–Si alloy in the heat-affected zone of a low-energy high-current electron beam and when exposed to combustion products during operation.

References for Chapter 5

1. Brandon, D. Microstructure of materials. Research and control methods / D. Brandon, W. Kaplan. - Moscow, Tekhnosfera, 2004 .- 384 p.
2. Utevsky, L.M. Diffraction electron microscopy in metal science / L.M. Utevsky. - M.: Metallurgiya, 1973. - 584 p.
3. Andrews, K. Electron diffraction patterns and their interpretation / K. Andrews, D. Dyson, S. Keone. - Moscow, Mir, 1971. - 256 p.
4. Kumar, C.S.S.R. (Ed.) Transmission Electron Microscopy Characterization of Nanomaterials / C.S.S.R. Kumar (Ed.). - New York: Springer, 2014 .- 717 p.
5. Barry Carter, C. Transmission Electron Microscopy / C. Barry Carter, B. David. - Berlin: Springer International Publishing, 2016 .- 518 p
6. Egerton, R.F. Physical Principles of Electron Microscopy / R.F. Egerton. - Berlin: Springer, 2016 .- 196 p.
7. Ivanov, Yu.F. Modification of structure and surface properties of hypoeutectic silumin by intense pulse electron beams / Yu.F. Ivanov, V.E. Gromov, S.V. Konovalov, D.V. Zagulyaev, E.A. Petrikova, A.P. Semin // Usp. Fiz. Met. - 2018 .- Vol. 19. - P. 195–222.
8. Zagulyaev, D.V. Study of the surface relief, structure and phase composition of the silumin composite layer obtained by the method of electric explosion alloying by Al-Y$_2$O$_3$ system / D.V. Zagulyaev, V.E. Gromov, Yu. F. Ivanov, E. A. Petrikova, A.D. Teresov, S.V. Konovalov, A.P. Semin // Journal of Physics: Conference Series. - 2018 .- Vol. 1115. - P. 1–7.
9. Mirzoev, F.Kh. Laser control processes in solids / F.Kh. Mirzoev, V.Ya. Panchenko, L.A. Shelepin // Usp. Fiz. - 1996. - Vol. 39. - P. 1–29.
10. Levchenko, E.B. The instability of surface waves in the inhomogeneously heated liquid / E.B. Levchenko, A.L. Chernyakov // Sov. Phys.–JETP. - 1981. – Vol. 54. - P. 102–105.
11. Mirzoev, F Kh. Evaporation-capillary instability in a deep vapor-gas cavity. Quantum Electronics. 1994. Vol. 24. - P. 138–140.
12. Bugaev, A.A. Thermocapillary phenomena and surface topography formation under the influence of picosecond laser pulses / A.A. Bugaev, V.A. Lukoshkin, V.A. Urpin, D.G. Yakovlev // Journal of Technical Physics. - 1988 .- T.58. - No. 5. - S. 908-914.
13. Sarychev, V. Model of nanostructure formation in Al – Si alloy at electron beam treatment / V. Sarychev, S. Nevskii, S. Konovalov, A. Granovskii, Y. Ivanov, V. Gromov // Materials Research Express. - 2019 .- Vol. 6. - P. 026540.
14. Papon, P. The Physics of Phase Transitions / P. Papon, J. Leblond, P. H. Meijer. - Berlin: Springer, 2002 .- 397 p.
15. Samokhin, A.A. Influence of evaporation on the melt behavior during laser interaction with metals / A.A. Samokhin // Soviet Journal of Quantum Electronics. - 1983. - Vol. 13. - P. 1347-1350.
16. Zong, X. High strain rate response of Ti-6.5Al-3.5Mo-1.5Zr-0.3 Si titanium alloy fabricated by laser additive manufacturing / X. Zong, Z. Li, J. Li, et al // Journal of Alloys and Compounds. - 2019 .- Vol. 781. - P. 47-55.
17. Qin, Y. Deep Modification of materials by thermal stress wave generated by irradiation of high-current pulsed electron beams / Y. Qin, C. Dong, Z.F. Song, S.Z. Hao, X.X. Me, J.A. Li, X.G. Wang, J.X. Zou, T. Grosdidier // Journal of Vacuum Science and Technology: A. - 2009 .- Vol. 23. - P. 430-435.

18. Robles Hernandez, F. C. Al–Si Alloys / F.C. Robles Hernandez, J.M. Herrera Ramírez, R. Mackay.– Berlin: Springer, 2017 .- 237 p.

19. Shahrooz, N. Semi-Solid Processing of Aluminium Alloys / N. Shahrooz, G. Reza. - Berlin: Springer, 2016 .- 363 p.

20. Stuwe, H.P. Shape instability of thin cylinders / H.P. Stuwe, O. Kolednik // Acta Metallurgica. - 1988. –Vol. 36 (7). - P. 1705–1708.

21. Ogris, E. On the silicon spheroidization in Al – Si alloys / E. Ogris, A. Wahlen, H. Luchinger, P.J. Uggowitzer // Journal of Light Metals. - 2002. –Vol. 2. - P. 263–269.

22. Grinfeld, M. Thermodynamic models of phase transformations and failure waves / M. Grinfeld // Wave Motion. - 2013 - Vol. 50 (7). - P. 1118–1126.

23. Grinfeld, M.A. Thermodynamic methods in the theory of heterogeneous substances / M.A. Grinfeld. - London: Longman, 1991 .- 399 p.

24. Buravova, S.N. Acceleration of Mass Transfer under Dynamic Loading / S.N. Buravova, E.V. Petrov // Russian Journal of Physical Chemistry B. - 2018 .- Vol. 12 (1). - P. 120–128.

25. Djahanbakhsh, M. Nanostructure formation and mechanical alloying in the wheel / rail contact area of high speed trains in comparison with other synthesis routes / M. Djahanbakhsh, W. Lojkowski, G. Bürkle, Yu.V. Ivanisenko, R.Z. Valiev, H.J. Fecht // Materials Science Forum. - 2001. - Vol. 360–362. P. 175–182

26. Gromov, V.E. Defect substructure change in 100-m differentially hardened rails in long-term operation / V.E. Gromov, A.A. Yuriev, Yu.F Ivanov. etc. // Materials Letters. - 2017 .- Vol. 209. - P. 224-227.

27. Konovalov, S. Mathematical Modeling of the Concentrated Energy Flow Effect on Metallic Materials / S. Konovalov, X. Chen, V. Sarychev, S. Nevskii, V. Gromov, M. Trtica // Metals. - 2017. - No. 7 (4). - P. 1–18.

28. Sarychev, V.D. Solution of niobium in iron during arc surfacing / V.D. Sarychev, B.B. Khaimzon, S.A. Nevskii // Steel in Translation. –2016.– V. 46 (8). - P. 563–566.

29. Liu, X. Heat-treatment induced defect formation in a-Al matrix in Sr-modified eutectic Al–Si alloy / X. Liu, B. Beausir, Y. Zhang et al. // Journal of Alloys and Compounds. - 2018 .- Vol. 730. - P. 208–218.

30. Sheng, Y. Application of high density electropulsing to improve the performance of metallic materials: mechanisms, microstructure and properties / Y. Sheng, Y. Hua, X. Wang, et al. // Materials. - 2018 .- V. 11 (2). - P. 185.

31. Kang, N. Microstructure and wear behavior of in-situ hypereutectic Al – high Si alloys produced by selective laser melting / N. Kang, P. Coddet, C. Chen, Y. Wang, H. Liao, C. Coddet / / Materials and Design. - 2016. - V. 99. - P. 120–126.

32. Brener, E.A. Dynamic instabilities of frictional sliding at a bimaterial interface / E.A. Brener, M. Weikamp, R. Spatschek, Y. Bar-Sinai, E. Bouchbinder // Journal of the Mechanics and Physics of Solids. - 2016. - Vol. 89. - P. 149–173.

33. Aldam, M. Nonmonotonicity of the Frictional Bimaterial Effect / M. Aldam, S. Xu, E. A. Brener, Y. Ben-Zion, E. Bouchbinder // Journal of Geophysical Research: Solid Earth. - 2017 .- Vol. 122. - P. 8270-8284.

34. 34. Sarychev, V. Model of nanostructure formation in Al – Si alloy at electron beam treatment / V. Sarychev, S. Nevskii, S. Konovalov, A. Granovskii, Y. Ivanov, V. Gromov // Materials Research Express. - 2019 .- Vol. 6 (2). - P. 026540.

35. Ivanov, Yu.F. Modification of structure and surface Properties of hypoeutectic silumin by intense pulse electron Beams / Yu.F. Ivanov, V.E. Gromov, S.V. Konovalov, D.V. Zagulyaev, E.A. Petrikova, A.P. Semin // Usp. Fiz. Met. - 2018 .- Vol. 19 (2). - P. 195-222.

36. Lavrentiev, M. A. Dynamic forms of loss of stability of elastic systems / M.A.

Lavrentiev, A.Yu. Ishlinsky // Doklady Physics. - 1949. - V. 5 (6). - P. 778-792.

37. Belyaev, A.K. The Lavrentiev – Ishlinsky problem at the initial stage of motion / A.K. Belyaev, N.F. Morozov, P.E. Tovstik, T.P. Tovstik // International Journal of Engineering Science. - 2016. - Vol. 98. - P. 92-98.

38. Kienzler, R. Theories of Plates and Shells / Ed. by. R. Kienzler. - Berlin: Springer, 2004 .- 238 p.

39. Eslami, M.R. Theory of Elasticity and Thermal Stresses / M.R. Eslami, R.B. Hetnarski, J. Ignaczak, N. Noda, N. Sumi, Y. Tanigawa. - Berlin: Springer, 2013 .- 789 p.

40. Sarychev, V. Disintegration mechanism of second phase particles under electron beams / V. Sarychev, S. Nevskii, S. Konovalov, A. Granovskii, V. Gromov // Materials Research Express. – 2019. – Vol. 6. – P. 106556.

6

Using intense pulsed electron beams for surface treatment of materials

Success in the field of generation of pulsed and continuous electron beams [1–3], development and development of appropriate equipment [4–8], numerous studies performed in the field of materials science of metals and alloys, cermet and ceramic materials treated with intense pulsed electron beams [5, 7, 8–16], prepared the basis for the use of such energy sources in various fields of industry, construction engineering and medicine [5, 6, 17, 18]. These directions continue to develop actively, as evidenced by numerous works [19–21], in which it is believed that electron beam processing is undoubtedly a promising technology that in some cases does not have an alternative at the moment.

To date, the following promising areas of use of the intense pulsed electron beams (IPEB) for metals and alloys and cermet materials have been identified:

- smoothing the surface, getting rid of surface microcracks, at the same time changing the structural phase state of the surface layer, which can be used, and in some cases used, to create high-performance finishing technologies for critical metal products of complex shape from titanium alloy Ti–6Al–4V and titanium [22–27]; steel of various classes [28–36]; carbide WC–10 wt.% Co [37]; aluminium [38].
- removal of microburrs formed during the manufacture of precision moulds (SKD11 steel) and biomedical products (Ti-6Al–4V alloy) [39].

- finishing surface treatment of molds and dies [32, 33, 35–37, 40].
- improvement of the functional properties of metallic biomaterials: stainless steel [34, 41–48], titanium and its alloys [22–24, 27, 31, 48, 49–54], alloys based on titanium nickelide with the shape memory effect [55], magnesium alloys [56–60],
- processing of medical devices and implants [22, 23, 61, 62],
- the formation of surface alloys for powerful electrodynamic systems [63, 64],
- improving the characteristics of the blades of aircraft engines and compressor blades [5, 65–67],
- the formation of thermal barrier coatings applied to the surface of the combustion chambers [68] and much more,
- surface treatment of rolling rails.

This section will briefly review examples of the use of IPEB (intense pulsed electron beam) metals and alloys, including Al–Si alloys used or recommended for use in industry.

It was shown in [38] that in the developed method of irradiating a large-area electron beam, a high energy density of the beam can be obtained without focusing it. Therefore, electron-beam surface treatment of a large area with a diameter of 60 mm can be used for instant melting and evaporation of the surface layer of the material, which expands the possibilities of highly effective surface treatment. However, if the preform has holes, the electron beam is concentrated on the input edge or the inner wall of the hole. Thus, it becomes difficult to smooth the bottom surface of the hole. Studies have shown that the effects of surface smoothing for a non-magnetic workpiece made of aluminium-based alloys can be enhanced by installing a magnetic block under the workpiece, since the electrons tend to concentrate with the magnetic field along the magnetic lines. In this study, the smoothing of the bottom surface of the hole was experimentally investigated by installing a magnetic block under the part when large area samples were irradiated with an electron beam. The expansion of the smoothing area of the lower surface was also tested by installing a through hole in the centre of the magnetic block. It was found that a wide area of the bottom surface of the hole can be smoothed out by irradiation of a large area by installing a magnetic block under the workpiece. The surface roughness decreased to less than 3.0 μm in the area of the surface of the bottom of the hole. When the diameter of the hole was 20 mm and the depth of the hole was less than 40 mm, a smoothing region of 10 mm diameter was obtained using a magnetic block. In addition, the smoothing area was sufficiently expanded by installing a through hole with an appropriate diameter in a magnetic field.

In [69], according to the analysis of studies published in [70–88], it was shown that the metals aluminium, magnesium and titanium, as well as alloys based on them, due to their low weight, are preferred for use in the spectrum of applications for which they are critical high performance and excellent combination of specific properties. The wider use of these materials in the aerospace, automotive and biomedical industries requires a significant improvement in their surface properties. Surface engineering is an economical and viable method for improving the surface properties of a material, such as hardness, wear resistance, and corrosion resistance, fatigue strength and oxidation resistance. Among the various methods of surface modification, high-energy processes based on the use of pulsed beams are very promising. Such methods include laser beams, plasma flows, powerful ion beams and electron beams.

Surface treatment by a pulsed electron beam has several advantages over other surface modification methods. A brief comparison of surface treatment by an intense pulsed electron beam (IPEB) with others used in surface modification processes is presented in Table 6.1 [69].

An analysis of the results given in Table 6.1 [69], as well as in many other publications [5–8], shows that methods of modifying the structure and surface properties of metals and alloys, cermet and ceramic materials based on the use of pulsed electron beams are promising and have significant advantages compared to other methods, which facilitates the processing of large areas, a significant penetration depth, along with low energy loss for ionization of the material. When an electron beam bombards the surface of the metal being processed, the material experiences the following transformations in layers arranged in series at different depths:

   (i) a surface molten layer,
   (ii) a heat affected zone; and
   (iii) a high-voltage zone arising under the influence of a shock wave formed as a result of bombardment of a material by an electron beam.

Figure 6.1 shows the results obtained by studying the methods of transmission electron diffraction microscopy of the cross section of irradiated samples of pre-hardened steel 45 (Fe–0.45C) [89, 90].

Table 6.1. Comparison of the surface treatment capabilities of a pulsed electron beam as compared with other methods used to modify the surface [69].

Other Surface Modification Methods	PEB Surface Treatment
1. Chemical Processes and Conversion Coatings	
Major Drawbacks	**Advantage of PEB Treatment**
(a) Use of toxic chemicals (e.g., chromate solutions for magnesium-based materials)	Absence of such issues
(b) Adhesion-related issues with substrates (i.e., interface in coatings and platings)	
2. Friction Stir Processing	
Major Drawbacks	**Advantage of PEB Treatment**
(a) Physical contact of tool with the surfaces, which distorts dimensional finish	No physical contact with surfaces of the chosen material
3. Pulsed Laser Beam Treatment	
Major Drawbacks	**Advantages of PEB Treatment**
(a) Low absorption depths (e.g., 0.02 μm, for laser pulse length ~10 to 50 ns)	(i) High absorption/melt depths (e.g., 2.6 μm, for an electron beam pulse length of 50 ns)
(b) Reflection up to 90% of incident energy (energy loss)	(ii) Only 5% to 10% of incident energy is reflected
(c) Limited to small surface area	(iii) Allows processing on large surface area (up to several 100 cm^2)
4. Pulsed Plasma Beam Treatment	
Major Drawbacks	**Advantage of PEB Treatment**
(a) Erosion of the material used for the electrode materials	Absence of such issues
(b) Deposition of electrode ions on the test sample	
(c) Formation and presence of toxic products	
5. Pulsed Ion Beam Treatment	
Major Drawbacks	**Advantages of PEB Treatment**
(i) Large loss in ionization	(i) Small loss in ionization
(ii) Small penetration depth (0.05 to1.0 mm, at 100 to 1000 keV kinetic energy)	(ii) Large penetration depth (10 to 500 mm, at 100 to 1000 keV kinetic energy)

Structural analysis of the cross section of the irradiated samples, performed by transmission electron diffraction microscopy, showed that, regardless of the number of pulses of the electron beam, the material has a depth-gradient structure consisting of several layers (Fig. 6.1). Upon single irradiation, a nanocrystalline layer with a thickness of ~0.1 μm is formed at the surface, consisting of grains α- (bcc solid solution based on iron) and γ- (fcc solid solution based on iron) phases in approximately equal proportions with an average grain size of ≈30 nm According to thermal calculations [91, 92], this layer was formed as a result of pulsed melting (melt lifetime is ~0.5 μs) and subsequent high-speed (up to ~1010 K/s) quenching from the melt. The velocity of the solidification front near the surface reaches ~5 m/s. A sublayer ~0.1 μm thick was formed under the nanocrystalline layer based on the α phase with an average grain size of ~200 nm. From thermal calculations it follows that this sublayer arose as a result of fast (~5 × 10^9 K/s) quenching from the slush state (a state characterized by the presence of islands of the solid phase located in the melt [93]). At depths of up to (0.3–0.5) μm, a sublayer with a mixed (α + γ) structure was

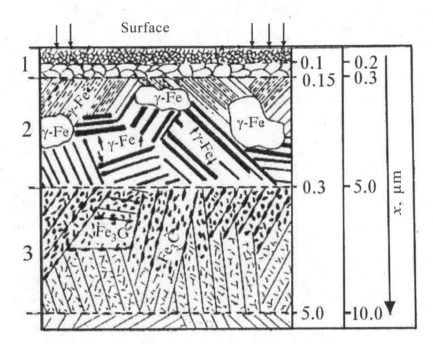

Fig. 6.1. Structural diagram of a sample of a pre-hardened steel 45 modified by an electron beam (pulse duration of an electron beam 0.8 μs, energy density of 2.2 J/cm²) [89, 90].

formed, preserving the morphology of the initial martensitic phase. At depths of ≥(0.3 0.5) μm, a ferrite–carbide mixture was observed, which is characteristic of martensite subjected to high-temperature tempering. At greater depths, the steel structure was similar to the initial one. With repeated (up to $n = 300$ pulses of exposure to an electron beam) irradiation, the characteristic layered structure of the material was preserved. However, a gradual increase in the initial temperature and the possible presence of pulses with an increased energy density [94] led to an increase in the thickness of the near-surface nanocrystalline ($\alpha + \gamma$) layer to ~0.2 μm and for the heat-affected zone to ~10 μm at ($n = 300$) [89 , 90].

With the right choice of process parameters, such as (*a*) accelerating voltage, (*b*) electron beam energy density, (*c*) number of pulses and (*d*) pulse duration, careful control and/or manipulation of the characteristics of the structural phase state and surface properties is possible. When IPEB surfaces of light metals (aluminium, magnesium and titanium), as well as alloys based on them, one or more microstructural transformations can occur [69]:

I)   selective enrichment with alloying elements of the surface layer,

II) the formation of a protective surface layer,

III) the formation of ultrafine (nanoscale) grain,

IV) dispersion of inclusions of the second phases,

V)   the formation of nonequilibrium phases due to rapid solidification,

VI) generation of a high density of dislocations due to shock wave action.

To improve the properties of the material and the durability of the products made from it, an important factor is the modification of the structure with the aim of forming submicron-sized grain (or subgrain structure) [95–101]. Surface melting and ultrafast hardening, which occurs during IPEB, make it possible to form a grain structure of the nanoscale range in the surface layer of the material. This process can be controlled by changing the parameters of the electron beam (energy density, duration and number of pulses) [7, 8, 11, 69].

For the Al–17.5Si alloy, the formation of a nanoscale grain structure was carried out by increasing the number of pulses of electron beam irradiation from 5, 15, 25 to 100 (Fig. 6.2 [72]).

It was found that after treatment with a pulsed electron beam, the surface molten layer of the Al–Si alloy under study contained a silicon-rich, aluminium-rich intermediate zone. Thin nanoscale silicon crystals of almost spherical shape were found in a silicon-rich zone surrounded by α-Al. Grains of free aluminium had a cellular structure, the dimensions of which were 100 nm. The intermediate zone contained ultrathin eutectic phases. As the number of pulses increases, the size of the silicon-rich zone increases; silicon gradually diffuses into the neighbouring zone, providing nucleation sites for the formation of ultrafine aluminium grains [72].

The results of studying the formation of a submicro-nanocrystalline structure in an Al–11Si–2Cu alloy irradiated by an intense pulsed electron beam with the following parameters: accelerated electron energy 17 keV, electron beam energy density (10, 15, 20, 25, 30, 35) J/cm^2, the pulse duration of the electron beam is 150 μs, the number of pulses is 3, and the pulse repetition rate is 0.3 s^{-1}, considered in a review article [102].

It was established [102] that processing an Al–Si alloy by a pulsed electron beam leads to the formation of a cellular type structure (Fig. 6.3). The thickness of the layer with the structure of cellular

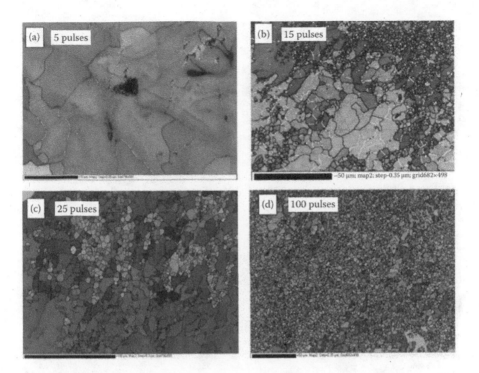

Fig. 6.2. Evolution of grain size in the aluminium-rich zone of the IPEB-treated Al–17.5 Si alloy for various number of pulses [72].

crystallization reaches 40 μm. The average cell size of the high-speed crystallization of the surface layer is 0.4 μm ± 0.11 μm. With a greater distance from the irradiation surface, the average crystallization cell sizes increase and reach 0.65 μm ± 0.22 μm at the lower boundary of the layer with the cellular structure.

The surface layer of an Al–Si alloy with a cell crystallization structure is characterized by the presence of lamellar eutectic grains (Fig. 6.4). The first eutectic grains are found in a layer located at a depth of ≈15 μm. As the distance from the irradiation surface increases, the relative content of eutectic grains increases. Eutectic grains are located in islands or interlayers between cells of high-speed crystallization of aluminium. The sizes of eutectic grains are close to the sizes of grains of a solid solution based on aluminium (crystallization cells). The transverse dimensions of the eutectic plates vary from 25 nm to 50 nm.

A layer in which aluminium and silicon is melted, and inclusions of foundry intermetallic compounds are preserved, is found at a depth of 50–70 μm (Fig. 6.5). In this case, the structure of cellular

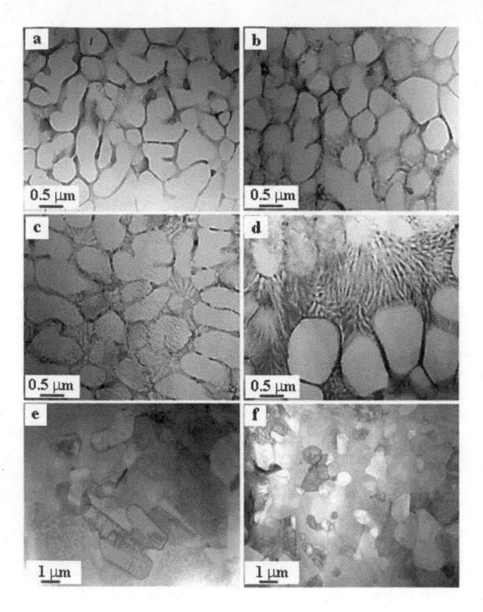

Fig. 6.3. TEM image of Al–Si alloy structure after electron beam treatment (E_s = 25 J/cm²): a – structure of 5 µm thick layer adjoining to the irradiation surface; b–f – structures of layers located at distances of X = 15, 30, 50, 120, 200 µm from the irradiation surface, respectively [102].

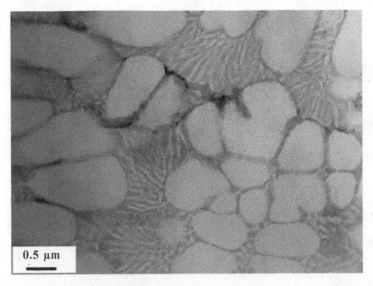

0.5 μm

Fig. 6.4. The structure of the alloy layer irradiated by an electron beam located at a depth of ≈30 μm (25 J/cm², 150 μs, 3 pulses) [102].

crystallization of aluminium and grains of the lamellar eutectic is formed. Inclusions of intermetallic compounds act as centres of cellular crystallization.

A layer in which only aluminium melts and primary inclusions of silicon and intermetallic compounds are present is detected at a distance of 80–90 μm from the irradiated surface. In this case, cells with high-speed crystallization of aluminium are observed in the structure. There are no grains of the lamellar eutectic of the submicron size (Fig. 6.6).

At a greater distance (100 μm or more) from the irradiated surface under this irradiation regime with an intense pulsed electron beam (25 J/cm², 150 μs, 3 pulses), the structure of the heat-affected zone is similar to the structure of the initial state of the material in terms of elemental and phase composition.

Microstructural modifications revealed in the alloy help to improve surface properties, namely, hardness, wear resistance, corrosion resistance, fatigue resistance, oxidation resistance and many other properties that are sensitive to the surface condition of the material. Due to this, the properties of light metals/alloys processed by pulsed electron beams in comparison with the untreated analogue are significantly increased.

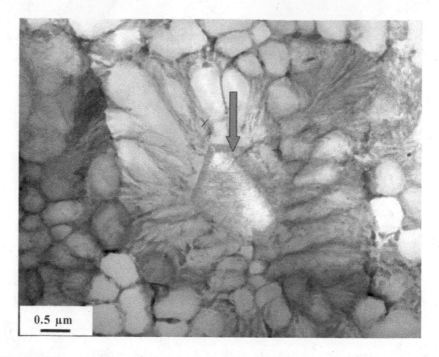

Fig. 6.5. The structure of the layer of material irradiated by an electron beam (25 J/cm², 150 μs, 3 pulses), located at a depth of ≈50 μm. The arrow indicates an intermetallic particle [102].

As noted above, the IPEB of the surface of metals and alloys is accompanied by high and ultrahigh (up to 10^9 K/s) cooling rates, which contributes to the formation of a structure with grains (cells of high-speed crystallization) of aluminium with sizes (250–600) nm and silicon crystals within 100 nm, having a rounded shape and evenly distributed in the modified volume of the alloy [103-105]. A methodology based on the use of this fact was developed in [106]. It was shown that the high cooling rate ($6 \cdot 10^2$ K/s) of the $AlSi_7Mg_{0.3}$ alloy immediately after applying pressure (about 100 MPa) during solidification of the casting with forced melt flow provides a significant grinding of eutectic silicon particles (up to 2–3 divisions of the Chai-Bäckerud six-stage scale) Such spheroidizing heat treatment not only increases the mechanical properties of castings, but also gives economic advantages due to the possibility of shortening the heating temperature for hardening.

It was shown in [107] that the formation of the structure of an Al–Si alloy with globular silicon inclusions has a significant effect on the tribological properties of the material. Antifriction Al–Si alloy with globular silicon (AGS) has been patented in Russia [108,

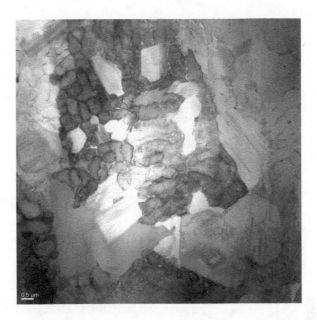

Fig. 6.6. The structure of the layer of material irradiated with an electron beam (25 J/cm², 150 μs, 3 pulses) located at a depth of ≈90 μm [102].

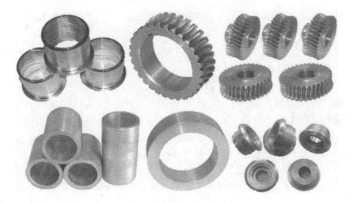

Fig. 6.7. Engineering components of antifriction Al–Si alloy with globular silicon.

109]. The ITM of the National Academy of Sciences of Belarus, produces continuous measured billets from AGSs with a diameter of up to 200 mm and a height of up to 250 mm; hollow billets with an outer diameter of 90 to 350 mm and a height of up to 200 mm; continuously cast rods with a diameter of 40 to 90 mm. The cost of procuring from AGS is 3 times less than that of a similar one from bronze. A general view of the parts made of such an alloy is shown in Fig. 6.7.

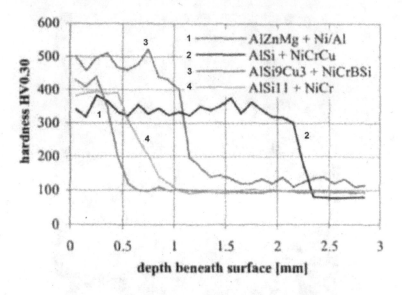

Fig. 6.8. Hardness profiles of aluminium -based alloys subjected to IPEB [113].

Delivery of blanks from AGS is carried out in accordance with TU BY 700002421.003-2011 to 60 enterprises of Belarus, Russia and Korea. The scope of applications for AGS blanks is sliding bearings, gears of worm gear wheels, bushings of balancers and hinges, liners of lathes of lathes and presses, bushings of differential gears and stuffing boxes, pistons of hydraulic cylinders, guide bushings and other details of technological equipment units. Thus, antifriction Al–Si alloy with globular silicon is a new antifriction material, which successfully replaces the heavier and more expensive serial antifriction bronzes.

In [110–113], it is noted that the industrial application of the IPEB of aluminium and alloy products based on it is especially promising in the automotive industry, namely, in a wide range of surface technologies. It is noted that electron beam treatment is successfully used to reduce the porosity of cast aluminium alloys, sputtering porous layers, and sintered materials. Good results on the successful use of electron beam processing were obtained by alloying, dispersing the structure or cladding of alloys based on iron, aluminium , titanium and magnesium. In Fig. 6.8 some results are

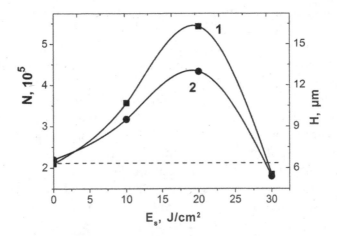

Fig. 6.9. Dependences of the number of cycles to failure N (curve 1) and thickness H of the surface layer (curve 2) on the energy density of the electron beam E_s, separated from the base by micropores. The dashed line marks the value of the fatigue life of steel in the initial state (before processing by the electron beam) [114]..

presented illustrating the effect of surface treatment by an electron beam on the mechanical properties of aluminium-based alloys.

A fundamentally important feature of the modification of the rolling surface of railway rails by low-energy, high-intensity electron beams is the absence of a pronounced interface between the modified layer and the volume of material, which determines good damping properties of the material under mechanical and thermal external influences, preventing premature nucleation and propagation of brittle microcracks from the surface into the bulk of the material leading to destruction [114–116].

Fatigue tests of steel revealed the dependence of the material durability on the electron beam energy density E_s (Fig. 6.9, curve 1). It is clearly seen that the maximum effect (an increase in the fatigue life of steel by ~2.5 times) is observed at $E_s = 20$ J/cm^2.

It was established that fatigue failure of steel is accompanied by the formation of a sublayer ~10 μm thick, at the interface of which micropores are located with the bulk of the material [115]. This circumstance suggests that fatigue failure of steel originates in the subsurface layer.

The process of pore formation is most pronounced when studying the fracture surface of steel treated with an electron beam at an electron beam energy density of 10 J/cm^2. The pore sizes in this case vary from 1 to 6 μm. In the steel treated with an electron beam at a higher beam energy density (20...30 J/cm²), the pore sizes are significantly smaller (0.3...1.0 μm). The lines formed by the pores are less pronounced, located at a certain distance from the irradiation surface, correlating with a change in the fatigue life of steel (Fig. 6.9, curve 2).

It has been shown that the predominant place for the formation of stress concentrators in an electron-irradiated steel beam is the interface between the high-speed crystallization layer and the heat-affected layer (bottom of the melt pool). It was established that the increase in the fatigue life of steel irradiated by an electron beam is due to the formation of a needle-shaped interface profile, which leads to dispersion of stress concentrators and contributes to a more uniform plastic flow in the substrate [114–116].

Thus, a brief analysis of the publication of the results of a study of the effect of pulsed electron beams on the structure and properties of the surface layer of metals and alloys, including Al-based alloys, allows us to conclude that the processing of industrial materials by pulsed electron beams will gradually take over to become the basis for future modification technologies for the purpose of engineering the surfaces of parts and products of a wide range of critical applications.

The scientific results of theoretical and experimental studies presented in the book allow us to create the foundations of the theory of the influence of electron–ion–plasma treatment on the structure, physical and mechanical properties of light alloys.

The developed methods for surface modification and analysis of structural changes make it possible to use the results of work for the development of scientific, scientific, technical and technological areas, while developing new technological solutions to increase the physical and mechanical properties of products from Al-based alloys for their further application in the automotive and space industries. The results of the work contribute to the development of subsequent development work on the introduction of electron–ion–plasma processing in the production cycle of metallurgical enterprises. This allowed us to develop recommendations and suggestions on the use of results in already developing innovative projects at enterprises of the Russian Federation. For example, the optimal PMPJ mode established in the work (the mass of the blown aluminium foil is 58.9 mg, the mass of the powdered sample of yttrium oxide is 88.3 mg, the discharge voltage of the capacitor bank is 2.6 kV) was used at LLC West 2002 Company for hardening and increasing the operational parameters of a worm wheel made of antifriction

Al–Si alloy and installed in the gearbox of a grinding and polishing machine 6ShP-100. Tests have shown that after PMPJ, the service life of the worm wheel increased by 3.5 times.

Remkomplekt Company established the feasibility of using IPEB to increase the life of the drive mechanisms used at the enterprise.

The recommended IPEB parameters are: accelerated electron energy 17 keV, electron beam pulse duration 150 µs, number of pulses 3, pulse repetition rate 0.3 s^{-1}, residual gas pressure (argon) in the working chamber of the installation $2 \cdot 10^{-2}$ Pa, energy density of the electron beam 25 J/cm^2.

References for Chapter 6

1.
2. Mesyats G.E., Proskurovsky D.I. Pulse electric discharge in vacuum. - Novosibirsk: Nauka, 1984. - 256 p.
3. Koval N.N., Ox E.M., Protasov Yu.S., Semashko N.N. Emission electronics. Publishing House of MSTU. N.E. Bauman, 2009 .-- 596 p. - (Ser. Electronics). (ISBN 978-5-70383347-6).
4. Proskurovsky D.I. Emission electronics. - Tomsk: Tomsk State University, 2010. - 288 p.
5. Koval N.N., Grigoriev S.V., Devyatkov V.N., Teresov A.D., Schanin P.M. Effects of intensified emission during the generation of a submillisecond low-energy electron beam in a plasma-cathode diode // IEEE Transactions on Plasma Science. 2009. V.37. No. 10. pp. 1890-1896.
6. Belov A.B., Bytsenko O.A., Kraynikov A.V. et al. High-current pulsed electron beams for aircraft engine building / Ed. A.S. Novikova, V.A. Shulova, V.I. Engelko. - M .: Deepak, 2012 .-- 292 p.
7. Ozur G.E., Proskurovsky D.I. Sources of low-energy high-current electron beams with a plasma cathode. - Novosibirsk: Nauka, 2018 .-- 176 p.
8. Evolution of the structure of the surface layer of steel subjected to electron-ion-plasma processing methods / Ed. N.N. Koval, Yu.F. Ivanova. - Tomsk: NTL Publishing House, 2016 .-- 298 p.
9. Electron-ion-plasma surface modification of non-ferrous metals and alloys / Ed. N.N. Koval, Yu.F. Ivanova. - Tomsk: NTL Publishing House, 2016 .-- 312 p.
10. Structure, phase composition, and properties of surface layers of titanium alloys after electroexplosive alloying and electron-beam processing / Ed. V.E. Gromova, Yu.F. Ivanova, E.A. Budovsky. - Novokuznetsk: Publishing house "Inter-Kuzbass", 2012. - 435 p.
11. Grishunin V.A., Gromov V.E., Ivanov Yu.F., Denisova Yu.A. Electron-beam modification of the structure and properties of steel. - Novokuznetsk; "Polygrafist", 2012. - 308 p.
12. Modification of the structure and properties of eutectic silumin by electron-ion-plasma treatment / A.P. Laskovnev, Yu.F. Ivanov, E.A. Petrikova et al .; under the editorship of A.P. Laskovneva. - Minsk: Belarus. Navuka, 2013 .-- 287 p.
13. Raikov S.V., Budovskikh E.A., Gromov V.E., Ivanov Yu.F., Vaschuk E.S. The formation of structural phase states and surface properties of titanium alloys during electroexplosive alloying and subsequent electron-beam processing. - Novokuznetsk: Inter-Kuzbass Publishing House, 2014. - 267 p.

14. Gromov V.E., Ivanov Yu.F., Vorobiev S.V., Konovalov S.V. Fatigue of steels modified by high intensity electron beams - Cambridge International Science Publishing. Cambridge - 2015 .- 272 p.

15. Sosnin K.V., Gromov V.E., Ivanov Yu.F. Structure, phase composition and properties of titanium after electro-explosive ·alloying with yttrium and electron-beam processing. - Novokuznetsk: Publishing house "Polygraphist", 2015. - 233 p.

16. Ivanov Yu.F., Gromov V.E., Konovalov S.V., Aksenova K.V. Fatigue of silumin modified by electron beam processing. - Novokuznetsk: Publishing house "Polygrafist", 2016. - 184 p.

17. Gromov V.E., Ivanov Yu.F., Konovalov S.V., Kormyshev V.E. Structure and properties of wear-resistant surfacing modified by electron-beam processing. - Novokuznetsk: Publishing. Centre of SibGIU, 2017 .-- 207 p.

18. Uno Y., Okada A., Uemura K., Raharjo P. Method for surface treating a die by electron beam irradiation and a die treated thereby. US patent No: US 7, 049, 539 B2, May 23, 2006.

19. Uemura K., Uehara S., Raharjo P., Proskurovsky D.I., Ozur G.E., Rotshtein V.P. Surface modification process on metal dentures, products produced thereby and the incorporated system therefore. US Patent No: US 6, 836, 531, B2, March 8, 2005.

20. Murray J.W., Kinnell P.K., Cannon A.H., Bailey B., Clare A.T. Surface finishing of intricate metal mold structures by large-area electron beam irradiation // Precision Engineering 37 (2013) 443-450.

21. Goriainov V., Cook R.B., Murray J.W., Walker J.C., Dunlop D.G., Clare A.T., Oreffo R.O.C. Human skeletal stem cell response to multiscale topography induced by large area electron beam irradiation surface treatment // Frontiers in Bioengineering and Biotechnology, 6 (2018) Article 91, 12 p. doi: 10.3389 / fbioe.2018.00091.

22. Manufacturing Techniques for Materials. Engineering and Engineered. Edited by T. S. Srivatsan, T. S. Sudarshan, K. Manigandan. Taylor & Francis Group, LLC. 2018. Chapter 10. 791 s.

23. Surface modification process on metal dentures, products produced thereby and the incorporated system therefore. US Patent No: US 6, 836, 531, B2, March 8, 2005.

24. P. Raharjo, Wada H., Nomura Y., Ozur G.E., Proskurovsky D.I., Rotshtein V.P., Uemura K., Pulsed electron beam technology for surface modification of dental materials. Proc. of 6th Intern. Conf. on Modification of Materials with Particle Beams and Plasma Flows, Tomsk, Russia, 2002, pp. 679- 682.

25. A. Okada, Y. Uno, N. Yabushita, K. Uemura, P. Raharjo. High efficient surface finishing of bio-titanium alloy by large-area electron beam irradiation, Journal of Materials Processing Technology 149 (2004) 506-511.

26. P. Raharjo, K. Uemura, A. Okada, Y. Uno. Application of large area electron beam irradiation for surface modification of implant materials, Proc. of 7th Intern. Conf. on Modification of Materials with Particle Beams and Plasma Flows, Tomsk, Russia, 2004, pp. 267-270.

27. A. Okada, Y. Uno, A. Iio, K. Fujiwara, K. Doi. New surface modification method of biotitanium alloy by electron-beam polising. Journ. of Advanced Mechanical Design, Systems and Manufacturing, vol. 2, No.4, 2008, pp. 695-700.

28. J. Tokunaga, T. Kojima, S. Kinuta, K. Wakabayashi, T. Nakamura, H. Yatani, T. Sohmura. Large-area electron beam irradiation for surface polishing of cast titanium. Dental Materials Journal (2009) 28 (5): 571-577.

29. A. Okada, Y. Uno, J.A. McGeough (1), K. Fujiwara, K. Doi, K. Uemura, S. Sano. Surface finishing of stainless steels for orthopedic surgical tools by large-area electron beam irradiation, CIRP Annals - Manufacturing Technology 57 (2008) 223–226

30. Y. Uno, A. Okada, K. Uemura, P. Raharjo. Method for surface treating a die by electron beam irradiation and a die treated thereby. US patent No: US 7, 049, 539 B2, May 23, 2006.

31. P. Raharjo, K. Uemura, A. Okada, Y. Uno. Application of large area electron beam irradiation for surface modification of metal dies, Proc. of 7th Intern. Conf. on Modification of Materials with Particle Beams and Plasma Flows, Tomsk, Russia, 2004, pp. 263-266.

32. Y. Uno, A. Okada, K. Uemura, P. Raharjo, T. Furukawa, K. Karato. High-efficiency finishing process for metal mold by large-area electron beam irradiation, Precision Engineering 29 (2005) 449–455.

33. Y. Uno, A. Okada, K. Uemura, P. Raharjo, S. Sano, Z. Yu, S. Mishima. A new polishing method of metal mold with large-area electron beam irradiation, Journal of Materials Processing Technology 187–188 (2007) 77–80.

34. A. Okada, Y. Okamoto, Y. Uno, K. Uemura. Improvement of surface characteristics for long life of metal molds by large-area EB irradiation. Journal of Materials Processing Technology 214 (2014) 1740-1748.

35. J.W. Murray, A.T. Clare Repair of EDM induced surface cracks by pulsed electron beam irradiation, Journal of Materials Processing Technology, 212 (2012) 2642–2651.

36. A. Selada, A. Manaia, M.T. Vieira, A.S. Pouzada. Effect of LBM and large-area EBM finishing on micro-injection molding surfaces // Int. J. Adv. Manuf. Technol., 2011, v. 52, pp. 171–182.

37. W. Murray, P.K. Kinnell, A.H. Cannon, B. Bailey, A.T. Clare, Surface finishing of intricate metal mold structures by large-area electron beam irradiation, Precision Engineering 37 (2013) 443-450.

38. A. Okada, R. Kitada, Y. Okamoto, Y. Uno. Surface modification of cemented carbide by EB polishing. CIRP Annals - Manufacturing Technology 60 (2011) 575-578.

39. T. Shinonaga, A. Okada, H. Liu, M. Kimura. Magnetic fixture for enhancement of smoothing effect by electron beam melting. Journal of Materials Processing Technology, 254 (2018) 229–237.

40. A. Okada, H. Yonehara, Y. Okamoto. Fundamental study on micro-deburring by large-area EB irradiation. Procedia CIRP 5 (2013) 19-24.

41. Y. Uno, A. Okada, K. Uemura, P. Raharjo. Method for surface treating a die by electron beam irradiation and a die treated thereby. US patent No: US 7, 049, 539 B2, May 23, 2006.

42. Goncharenko I.M., Itin V.I., Isichenko S.V., Lykov S.V., Markov A.B., Nalesnik O.I., Ozur G.E., Proskurovsky D.I., Rotstein V.P. Increasing the corrosion resistance of steel 12Cr18Ni10Ti during processing with a low-energy high-current electron beam // Zashchita metallov, 1993, v. 29, no. 6, p. 932-937.

43. G.S. Eklund, Initiation of pitting at sulfide inclusions in stainless steel, J. Electrochem. Soc. 121 (1974) 467–473.

44. T. Suter, H. Böhni, A new microelectrochemical method to study pit initiation on stainless steels, Electrochim. Acta 42 (20-22) (1997) 3275-3280.

45. Manning, C.E. Lyman and D.J. Duquette, A STEM Examination of the localized corrosion behavior of a duplex stainless steel, Corrosion 36 (5) (1980) 246-251.

46. A.J. Sedriks, Effects of alloy composition and microstructure on the passivity of stainless steels, Corrosion 42 (1986) 376–389.

47. A.V. Batrakov, A.B. Markov, G.E. Ozur, D.I. Proskurovsky, V.P. Rotshtein, Effect of pulsed electron-beam treatment of electrodes on the electric strength of the vacuum insulation, IEEE Transactions on Dielectrics and Electrical Insulation 2 (2) (1995)

237-242. doi: 10.1117 / 12.174677.

48. K. Zhang, J. Zou, T. Grosdidier, C. Dong, D. Yang, Improved pitting corrosion resistance of AISI 316L stainless steel treated by high current pulsed electron beam, Surface & Coatings Technology 201 (2006) 1393–1400 .

49. Rothstein V.P., Gunzel R., Markov A.B., Proskurovsky D.I., Fam M.T., Richter E., Shulov V.A. Surface modification of VT6 titanium alloy with a low-energy high-current electron beam at elevated initial temperatures. Physics and chemistry of materials processing. 1 (2006) 62-72.

50. X.D. Zhang, S.Z. Hao, X.N. Li, C. Dong, T. Grosdidier, Surface modification of pure titanium by pulsed electron beam, Applied Surface Science 257 (2011) 5899-5902.

51. G. Guo, G. Tang, X. Ma, M. Sun, G.E. Ozur, Effect of high current pulsed electron beam irradiation on wear and corrosion resistance of Ti6Al4V, Surf. Coat. Technol. 229 (2013) 140-145.

52. A. Balyanov, J. Kutnyakova, N.A. Amirkhanova, V.V. Stolyarov, R.Z. Valiev, X.Z. Liao, Y.H. Zhao, Y.B. Jiang, H.F. Xu, T.C. Lowe, Y.T. Zhu, Corrosion resistance of ultra finegrained Ti, Scr. Mat. 51 (2004) 225-229.

53. J.C. Walker, J.W. Murray, M. Nie, R.B. Cook, A.T. Clare, The effect of large-area pulsed electron beam melting on the corrosion and microstructure of a Ti6Al4V alloy, Appl. Surf Sci. 311 (2014) 534-540.

54. V. Goriainov, R.B. Cook, J.W. Murray, J.C. Walker, D.G. Dunlop, A.T. Clare, R.O.C. Oreffo, Human skeletal stem cell response to multiscale topography induced by large area electron beam irradiation surface treatment, Frontiers in Bioengineering and Biotechnology, 6 (2018) Article 91, 12 p. doi: 10.3389 / fbioe.2018.00091

55. B.W. Stuart, J.W. Murray, D. M. Grant, Two step porosification of biomimetic thin-film hydroxyapatite / alpha-tri calcium phosphate coatings by pulsed electron beam irradiation, Sci. Rep. 8 (2018) 14530. https://doi.org/10.1038/s41598-018-32612-x.

56. K.M. Zhang, D.Z. Yang, J.X. Zou, T. Grosdidier, C. Dong, Improved in vitro corrosion resistance of a NiTi alloy by high current pulsed electron beam treatment, Surface & Coatings Technology 201 (2006) 3096-3102.

57. S. Hao, B. Gao, A. Wu, J. Zou, Y. Qin, C. Dong, J. An, Q. Guan, Surface modification of steels and magnesium alloy by high current pulsed electron beam, Nuclear Instruments and Methods in Physics Research B 240 (2005) 646–652. AZ91HP.

58. B. Gao, S. Hao, J. Zou, W. Wu, G. Tu, and C. Dong, Effect of high current pulsed electron beam treatment on surface microstructure and wear and corrosion resistance of an AZ91HP magnesium alloy, Surf . Coat. Technol. 201 (14) (2007) 6297-6303.

59. M.C. Li, S.Z. Hao, H. Wen, R.F. Huang, Surface composite nanostructures of AZ91 magnesium alloy induced by high current pulsed electron beam treatment, Applied Surface Science 303 (2014) 350-353. https://doi.org/10.1016/j.apsusc.2014.03.03.00440.

60. S. Hao, M. Li, Producing nano-grained and Al-enriched surface microstructure on AZ91 magnesium alloy by high current pulsed electron beam treatment, Nuclear Instruments and Methods in Physics Research Section B: Beam Interactions with Materials and Atoms 375 (2016) 1-4. https://doi.org/10.1016/j.nimb.2016.03.03.035.

61. Y.R. Liu, K.M. Zhang, J.X. Zou, D.K. Liu, T.C. Zhang, Effect of the high current pulsed electron beam treatment on the surface microstructure and corrosion resistance of a Mg-4Sm alloy, Journal of Alloys and Compounds 741 (2018) 65-75.

62. K. Uemura. Thin film, material modification and application technologies. Session report. 16th Simp. on High Current Electronics & 10th Conf. on Materials Modification. Tomsk, Russia, Sept. 24, 2010.

63. T. Mori, A. Okawa, T. Tsubouchi, K. Uemura. Manufacturing method for medical

equipment for reducing platelet adhesion on a surface in contact with blood. Patent US 8,997,349 B2. Apr. 7, 2015. Prior Publ. Data: US 2013 / 0230422A1. Sept. 5, 2013.

64.   A.B. Markov, E.V. Yakovlev, V.I. Petrov. Formation of surface alloys with a low-energy high-current electron beam for improving the high-voltage hold-off of copper electrodes. IEEE Trans. Plasma Sci., 2013, v. 41, No. 8, pp. 2177–2182. DOI: 10.1109 / TPS.2013.2254501.

65.   A.V. Batrakov, S.A. Onischenko, I.K. Kurkan, V.V. Rostov, E.V. Yakovlev, E.V. Nefedtsev, R.V. Tsygankov, Comparative study of breakdown strength of vacuum insulation in gaps with electron-beam polished electrodes under pulsed DC and microwave electric fields // Proc. 28th Int. Symp on Discharges & Electrical Insulation in Vacuum, Greifswald, Germany, September 23–28, 2018, v. 1, pp. 77–80.

66.   Proskurovsky D.I., Rotshtein V.P., Ozur G.E. Use of low-energy, high-current electron beams for surface treatment of materials // Surf. Coat. Technol. - 1997. - Vol. 96, N1. - P.115-122.

67.   Proskurovsky D.I., Rotshtein V.P., Ozur G.E. et al. Pulsed electron-beam technology for surface modification of metallic materials // J. Vac. Sci. Technol. - 1998. - Vol. A16 (4), N1. - P.2480-2488.

68.   Yagodkin Yu.D., Pastukhov KM, Kuznetsov S.I. et al. The effect of irradiation by a powerful electron beam on surface topography and the physicochemical state of the surface layer of heat-resistant alloys // Physics and Chemistry of Materials Processing. - 1995. - Vol. 5. - S. 111-119.

69.   Cai J., Guan Q., Xu X., Lu J., Wang Z., Han Z. Thermal cycling bechavior of thermal barrier coatings with MCrAlY bond coat irradiated by high-current pulsed electron beam // ACS Appl. Mater. Interfaces - 2016. - Vol. 47, N8. - P. 32541-32556.

70.   Subramanian Jayalakshmi, Ramachandra Arvind Singh, Sergey Konovalov, Xizhang Chen, and T. S. Srivatsan. Overview of Pulsed Electron Beam Treatment of Light Metals: Advantages and applications "Manufacturing Techniques for Materials. Engineering and Engineered. " Edited by T. S. Srivatsan T. S. Sudarshan K. Manigandan. Taylor & Francis Group, LLC. 2018. Chapter 11. S. 327-366.

71.   An J., X. X. Shen, Y. Lu, Y. B. Liu, R. G. Li, C. M. Chen and M. J. Zhang (2006). Influence of high current pulsed electron beam treatment on the tribological properties of Al – Si – Pb alloy. Surface & Coatings Technology 200: 5590–5597.

72.   Dong, C., A. Wu, S. Hao, J. Zou, Z. Liu, P. Zhong, A. Zhang, T. Xu, J. Chen, J. Xu, Q. Liu and Z. Zhou (2003). Surface treatment by high current pulsed electron beam. Surface and Coatings Technology 163–164: 620–624.

73.   Gao, B., L. Hu, S. W. Li, Y. Hao, Y. D. Zhang, G. F. Tu and T. Grosdidier (2015). Study on the nanostructure formation mechanism on the hypereutectic Al – 17 Si alloy induced by pulsed electron beam. Applied Surface Science 346: 147–157.

74.   Gao, B., S. Hao, J. Zou, T. Grosdidier, L. Jiang, J. Zhou and C. Dong (2005). High current pulsed electron treatment of AZ31 Mg alloy. Journal of Vacuum Science & Technology A 23 (6): 1548-1553.

75.   Gao, B., S. Hao, J. Zou, W. Wu, G. Tu and C. Dong (2007). Effect of high current pulsed electron beam treatment on surface microstructure and wear and corrosion resistance of an AZ91HP magnesium alloy. Surface & Coatings Technology 201: 6297-6303.

76.   Gromov, V. E., Yu. F. Ivanov, A. M. Glezerd, S. V. Konovalov and K. V. Alsaraeva (2015). Structural evolution of silumin treated with a high intensity pulse electron beam and subsequent fatigue loading up to failure. Bulletin of the Russian Academy of Sciences-Physics 79 (9): 1169–1172.

77. Hao, S., S. Yao, J. Guan, A. Wu, P. Zhong and C. Dong (2001). Surface treatment of aluminium by high current pulsed electron beam. Current Applied Physics 1 (2001): 203–208.

78. Hao, S. Z., Y. Qin, X. X. Mei, B. Gao, J. X. Zuo, Q. F. Guan, C. Dong and Q. Y. Zhang (2007). Fundamentals and applications of material modification by intense pulsed beams. Surface & Coatings Technology 201: 8588–8595.

79. Hao, Y., B. Gao, G. F. Tu, S. W. Li, C. Dong and Z. G. Zhang (2011a). Improved wear resistance of Al–15 Si alloy with a high current pulsed electron beam treatment. Nuclear Instruments and Methods in Physics Research B 269: 1499-1505.

80. Hao, Y., B. Gao, G. F. Tu, S. W. Li, S. Z. Hao and C. Dong (2011b). Surface modification of Al – 20 Si alloy by high current pulsed electron beam. Applied Surface Science 257: 3913–3919.

81. Yu. F. Ivanov, K. V. Aksenova, V. E. Gromov, S. V. Konovalov, and E. A. Petrikova (2016). An increase in fatigue service life of eutectic silumin by electron beam treatment. Russian Journal of Non-Ferrous Metals 57 (3): 236–242.

82. Karlsruhe Institute of Technology (2014). Surface modification of materials using pulsed electron beams. https://www.ihm.kit.edu/english/300.php.

83. Kim, J. S., W. J. Lee, and H. W. Park (2016). The state of the art in the electron beam manufacturing processes. International Journal of Precision Engineering and Manufacturing 17 (11): 1575-1585.

84. Konovalov, S. V., K. V. Alsaraeva, V. E. Gromov and Yu. F. Ivanov (2015a). Fatigue life of silumin irradiated by high intensity pulsed electron beam. IOP Conference Series: Materials Science and Engineering 91: 012029.

85. Konovalov, S. V., K. V. Alsaraeva, V. E. Gromov, Yu. F. Ivanov and O. A. Semina (2016). Fatigue variation of surface properties of silumin resulting to electron-beam treatment. IOP Conference Series: Materials Science and Engineering 110: 012012.

86. Konovalov, S. V., K. V. Alsaraeva, V. E. Gromov and Y. F. Ivanov (2015b). Structure-phase states of silumin surface layer after electron beam and high cycle fatigue. The 12th International Conference on Gas Discharge Plasmas and Their Applications. Journal of Physics: Conference Series 652: 012028.

87. Mueller, G., V. Engelko, A. Weisenburger and A. Heinzel (2005). Surface alloying by pulsed intense electron beams. Vacuum 77: 469–474.

88. Total Materia (2004b). Aircraft and aerospace applications: Part 1 & 2. http://www. totalmateria.com/Article96 .htm (Last Accessed: June 15, 2017).

89. Uno, Y., A. Okada, K. Uemura, P. Raharjo, T. Furukawa and K. Karato (2005). High-efficiency finishing process for metal mold by large-area electron beam irradiation. Precision Engineering 29 (4): 449–455

90. Ivanov Yu.F., Itin V.I., Lykov S.V., Markov A.B., Rotshtein V.P., Tukhfatullin A.A., Wild N.P. Structural analysis of the heat-affected zone of steel 45 processed by a low-energy high-current electron beam // Physics of Metals and Metallurgy. - 1993. - T. 75, no. 5. - S. 103–112.

91. Ivanov Yu.F., Itin V.I., Lykov S.V., Markov A.B., Mesyats G.A., Ozur G.E., Proskurovsky D.I., Rotshtein V.P., Tukhfatullin A.A. Phase and structural transformations in steel under the influence of a low-energy high-current electron beam // Metals. - 1993. - Vol. 3. - S. 130–140.

92. Markov A.B., Rotshhtein V.P. Calculation and experimental determination of dimensions of hardening and tempering zones in quenched U7A steel irradiated with a pulsed electron beam // NIM B. - 1996.- V.132. - P.79-86.

93. Markov A.B., Rotshtein V.P. Calculation and experimental determination of the size of hardening and tempering zones in hardened steel U7A irradiated with a pulsed

electron beam // Surface. X-ray, synchrotron and neutron studies. - 1998.- No. 4.-P.83-89.

94.   Markov A.V., Proskurovsky D.I., Rotshtein V.P. The formation of the heat-affected zone in iron and steel 45 when exposed to low-energy high-current electron beams. - Tomsk: Ed. TSC SB RAS, 1993 .-- 63 p.

95.   Markov A.B., Rotshtein V.P. The mechanism of increasing the thickness of the heat-affected zone during pulse-periodic processing of the target by an electron beam // Thermophysics of high temperatures. -2000. –T.38, No. 1. –S.19-23.

96.   Valiev R.Z., Alexandrov I.V. Nanostructured materials obtained by intense plastic deformation. - M .: Logos, 2000 .-- 272 p.

97.   Noskova N.I., Mulukov R.R. Submicrocrystalline and nanocrystalline metals and alloys. - Yekaterinburg: Ural Branch of the Russian Academy of Sciences, 2003 .- 279 p.

98.   Valiev R.Z., Alexandrov I.V. Bulk nanostructured metallic materials: production, structure and properties - Moscow: Academkniga, 2007. - 397 p. ISBN 978-5-94628-217.

99.   Structure and properties of vtnfls at different energy effects and treatment technologies / Edit. by V.A. Klimenov and V.A. Starenchenko. - Switzerland. - Trans Tech Publications Ltd, 2014 .-- 312 p.

100.   Kozlov E.V., Glezer A.M., Koneva N.A., Popova N.A., Kurzina I.A. Fundamentals of plastic deformation of nanostructured materials / Ed. A.M. Glaser. - M .: Fizmatlit, 2016 .-- 304 s.

101.   Valiev R.Z., Estrin Y., Horita Z., Langdon T.G., Zehetbauer M.J., and Zhu Y. Producing Bulk Ultrafine-Grained Materials by Severe Plastic Deformation: Ten Years Later // JOM. - 2016. - Vol. 68, No. 4. - R. 1216-1226. DOI: 10.1007 / s11837-016-1820-6.

102.   High technology in projects of the Russian Science Foundation. Siberia / Ed. S.G. Psakhie, Yu.P. Sharkeeva. - Tomsk: NTL Publishing House, 2017 .-- 428 p.

103.   Yu. F. Ivanov, D. V. Zagulyaev, S. A. Nevskii, V. E. Gromov, V. D. Sarychev, and A. P. Semin. Microstructure and Properties of Hypoeutectic Silumin Treated by High-Current Pulsed Electron Beams // Progress in Physics of Metals, 2019, vol. 20, pp. 451-490 (https://doi.org/10.15407/ufm.20.03.451).

104.   D. Zaguliaev, S. Konovalov, Yu. Ivanov, V. Gromov, and E. Petrikova Microstructure and mechanical properties of doped and electron-beam treated surface of hypereutectic Al-11.1% Si alloy // Journal of Materials Research and Technology, 2019, vol. 8, Issue 5, pp. 3835-3842. (https://doi.org/10.1016/j.jmrt.2019.06.045).

105.   V. Sarychev, S.Nevskii, et al., Model of nanostructure formation in Al – Si alloy at electron beam treatment // Mater. Res. Express 6 (2019) 026540, p. 1-14. (http://doi.org/10.1088/2053-1591/aaec1f).

106.   Yurii Ivanov, Victor Gromov, Dmitrii Zaguliaev, Alexander Gleze, Roman Sundeev, Yulia Rubannikova, Alexander Semin. Modification of surface layer of hypoeutectic silumin by electroexplosion alloying followed by electron beam processing // Materials Letters 253 (2019) 55–58 (https://doi.org/10.1016/j.matlet.2019.05.148).

107.   Stanèek L., Vanko B., and A. I. Batyshev. Structure and properties of silumin castings solidified under pressure after heat treatment // Metal Science and Heat Treatment, Vol. 56, Nos. 3 - 4, July, 2014, pp. 179-202.

108.   E. I. Marukovich, V. Yu. Stetsenko, A. P. Gutev. Manufacture and use of silumin with globular silicon // Casting and Metallurgy. - 2017 .-- T. 2 (87). - S. 15-19.

109.   Marukovich E. I., Stetsenko V. Ju. Sposob ohlazhdenija kristallizatora [Method of cooling the mold]. Patent RU, no. 2342200, 2008.

110.   Stetsenko V. Yu., Marukovich E. I. Antifrikcionnyj splav na osnove aljuminija [Aluminium -based antifriction alloy]. Patent RU, no. 25004595, 2014.

111.   R. Zenker, Electron beam surface treatment, industrial applications and prospects // Surf. Eng. 1996, 12 (4), pp. 9-12.

112.   Zenker R. A New electron beam surface technology / R. Zenker, B. Furchheim // International Heat Treating Conference - Equipment and Processes, Schaumburg, 1994, pp. 299-301.

113.   R. Zenker, N. Frenkler, T. Ptaszek, Neuent-wicklungen auf dem Gebiet der EB Rand-schichtbehandlung. In: HTM 54 (1999), 3, pp. 143-149.

114.   Zenker Rolf, Buchwalder Anja. Application of Electron Beam Surface Technologies in the Automotive Industry // Transactions of materials and heat treatment proceedings of the 14 ™ ifhtse congress. - October 2004. - Vol.25. - No.5. - R. 573-578.

115.   Ivanov, Yu. F. Structure of the surface layer and fatigue life of a rail steel irradiated with a high-intensity electron beam / Yu. F. Ivanov, V. E. Gromov, V. A. Grishunin [et al.]. // Physical Mesomechanics. - 2013 .-- T.16. - No. 2. - S. 47 - 53.

116.   Ivanov, Yu. F. Electron-beam processing of rail steel: phase composition, structure, fatigue life / Yu. F. Ivanov, V. E. Gromov, V. A. Grishunin [et al.] // Questions of materials science. - 2013. - No. 1 (73). - S. 20 - 30.

117.   Gromov, V.E. Scale levels of structural-phase states and fatigue life of rail steel after electron-beam processing / V. E. Gromov, Yu. F. Ivanov, V. A. Grishunin [et al.] // Successes in metal physics. 2013 .-- v. 14. - No. 1. - S. 67 - 80.

Index